Praise for *Healthy Bee, Sick Bee*

'If you want to learn about popping bee penises (n't), and where not to get stung, then this is th-   his book contains much more than a stri   il Lester not only shows us his love of honey   ge of everything that plagues them. I can easily   becoming the go-to-book for anyone interested in he   eir demons, large and small. Highly recommended.' —Prof Ma   e Beekman, School of Life and Environmental Sciences, University of Sydney

'Lester's book is an impeccably researched analysis of the health of the honey bees of New Zealand by one of the world's greatest bee pathologists. Lester pulls no punches when discussing controversial topics like the effects of neonicotinoids on bees, or the (un)likelihood of eradicating AFB from New Zealand. Packed with anecdote and cautionary tales, the story is told with immense style. This will be a valuable resource for beekeeper and scientist alike.' —Prof Ben Oldroyd, School of Life and Environmental Sciences, University of Sydney

Praise for *The Vulgar Wasp*

'Lester cleverly weaves facts and figures on the astonishing science of this little-loved insect into a text that's tickled with memorable anecdotes and personable insights. If you thought wasps were pointless, boring and unimportant, think again.' —Dr Seirian Sumner, Reader in Behavioural Ecology at University College London

'*The Vulgar Wasp* is more than just interesting and instructive; it's a delight to read. It's packed full of up-to-date information on invasive wasps in New Zealand and elsewhere, as well as in their native range – all presented in an engaging, humorous and informative manner. Everything you might want to know about the life cycle of wasps, methods of control, and the environmental, social and economic costs of these pesky invertebrates can be found in this enormously readable book.' —Dr Andrea Byrom, Director of New Zealand's Biological Heritage National Science Challenge

HEALTHY BEE, SICK BEE

The Influence of Parasites, Pathogens, Predators and Pesticides on Honey Bees

PHIL LESTER

 TE HERENGA WAKA
UNIVERSITY PRESS

Te Herenga Waka University Press
PO Box 600 Wellington
teherengawakapress.co.nz

A catalogue record for this book is available from
the National Library of New Zealand.

ISBN 9781776564057

Printed by Markono Print Media Pte Ltd, Singapore

'The happiness of the bee and the dolphin is to exist. For man it is to know that and to wonder at it.'

—Jacques-Yves Cousteau

CONTENTS

Honey bees are essential to our agricultural and horticultural industries. Pasture crops like clover and alfalfa need bee pollination. *Photo: Phil Lester*

INTRODUCTION

What kills honey bees?

'If the bee disappeared off the face of the Earth, man would only have four years left to live.' That quote is widely attributed to Albert Einstein. It's the sort of substantial and big-picture statement that this impressive man could have made, but there is no actual evidence that he said it.

Without a doubt, however, bees and other pollinators are extraordinarily important. We depend on pollination for much of our food. A research group from Europe estimated that the economic value of pollination worldwide amounted to €153 billion. Vegetables and fruits were the leading benefactors of insect pollination, followed by edible oil crops, stimulants, nuts and spices. Beef and dairy production is heavily reliant on forage pollination by insects. Pollinators have been estimated to contribute to 9.5% of the total value of the production of human food worldwide.[1] A total of 87 of the 115 leading global food crops are dependent upon animal pollination, and those animals are mostly bees.

Of all the bee pollinators, the western honey bee or European honey bee, *Apis mellifera*, is the most economically valuable in cropping systems around the world. Without honey bees, the yields of many fruit, nut and seed crops would drop by more than 90%.[2] Perhaps the most extreme example of our reliance on honey bees is in almond production. Nearly 80% of the world's almonds are produced in the Central Valley of California each year, and almonds are reliant on insect pollination. It has been estimated that over 60% of the commercially managed honey beehives in the United States are transported from across the country each year to these almond groves. Almond growers need honey bees; their economic livelihood is in jeopardy without them. From the *Los Angeles Times* in 2016: 'Without bees, there can be no almonds. In fact, each of California's nearly 1 million acres of almond orchards requires two hives. But California beekeepers have only a quarter of the needed hives. As almond acreage has exploded and bees have been in some kind of crazy death spiral, growers have been in a mild state of panic over where to find enough little pollinators.'[3]

Even the world's superpowers recognise the economic importance of honey bees. President Obama had beehive and pollinators' garden installed on the South Lawn of the White House in 2015. These 35,000 bees are probably the only bees anywhere protected by the Secret Service. 'I do care about bees – and we're going to fix them!' said President Obama. At that time the he was launching a plan to reduce honey bee overwintering colony losses to no more than 15% within 10 years.[4] Given the high rate of overwintering loss at that time in the US of 27%,[5] the goal of losing just 15% represents hundreds of thousands of hives surviving when normally they would die. Not to be outdone by the United States, New Zealand's current prime minister also keeps bees on government grounds. In order to mark the beginning of Bee Aware Month in 2019, the current minister of agriculture, Damien O'Connor, inspected the prime minister's hives and told a group of young beekeepers from Te Aro

A worker (female) bee. Note the large hind legs, which are well-suited to transporting pollen. *Photo: Phil Lester*

School that bees were the most important animal in the world and needed our protection. 'Without bees we wouldn't have pollination, and without pollination we wouldn't have food. If we look after the bees then they can look after pollination.'[6]

We are so fascinated by bees that we have sent them into space. In 1982 a space shuttle carried common house flies, velvetbean caterpillar moths, and 14 honey bees into orbit. The poor bees had their stingers clipped to reduce the danger to the crew. In space, the bees tended to float a lot. Near the end of the trip, one observant and astute astronaut noted that 'the bees have all gotten stationary'.[7] All 14 of the honey bees had died. It seems that they had been given inadequate food, though it really sounded like they had been starved and/or bored to death before the shuttle returned to Earth.[8]

In 1984, in an experiment designed by a high-school student, two complete colonies of honey bees, each with around 3400 workers and a queen, were blasted into orbit on the space shuttle *Discovery*. For some strange reason NASA officials deemed dead bees to be the primary concern for human health on

A drone (male) bee. The drone is much stockier than the worker, with large thoracic muscles powering the wings, and large eyes which help the bee to spot virgin queens in drone congregation areas. *Photo: Phil Lester*

this trip, so the enclosure containing the bees was given a filtration system to address concerns about hazards posed by dead bees. All 6800 bees on board were allowed to keep their stingers, and attempts were made to provide nicer in-flight meals for them and a degree of stimulation. The bees needed a day

Top: Astronaut C. Gordon Fullerton with insects, March 1982. Moths, flies and 14 bees were studied during an eight-day flight on the space shuttle *Columbia*. The poor bees had their stingers clipped prior to take-off.
Below: Astronaut James D. van Hoften inspects the Bee Enclosure Module with its colony of 3400 honey bees. By the end of their seven-day orbit in April 1984, the bees had adjusted to their new environment. *Photos: NASA Library*

or two in space to get used to zero gravity. By the end of the seven-day orbit, they had adjusted to their new environment and 'showed complete adaptation to microgravity'. They flew from place to place. They built wax comb and the queens laid eggs. Crew members noted that bees were able to learn and adjust their flight behaviour to efficiently move around the colony. All but 350 of the 6800 bees survived the trip.[9] No doubt to the great relief of NASA, the 350 deceased bees didn't seem to have affected astronaut health. This rate of bee (and astronaut) mortality would be about what you'd expect from bee colonies of this size on Earth over a week-long period.

The entire world seems to love honey bees. Our love of this insect is a little ironic given that their stings probably result in more human deaths than any other animal in many countries, including New Zealand. (Malaria-carrying mosquitoes take that prize in other countries and continents.)

Beekeeper Charlie Brandts works with the beehive on the South Grounds of the White House, May 2015. *Photo: The White House / Pete Souza*

A long history of bee declines and colony collapse

One reason for our long-held love of honey bees is our sweet tooth. Until only a few centuries ago, honey was the only sweetener available. Honey is a favourite on morning toast around the globe. It stores well and has a huge range of uses. Because we love honey so much, we've been selecting honey bees for traits that we prefer, such as highly productive, gentle bees that are less inclined to sting the beekeepers who steal the results of their hard work. The bees we see today behave differently and probably look different from those observed even a few hundred years ago.

We've been writing about and recording information on bees for thousands of years. Among those writings is a long history of honey bee declines and collapses at a national scale. One of our best records is still the veterinary surgeon George Fleming's 1871 book *Animal Plagues: Their History, Nature and Prevention*. It includes the first mention of a widespread 'mortality of bees' in Ireland in the year 950. Fleming described how bees were a source of wealth to the Irish people. We have no idea what caused the decline of these Irish bees, but at around the same time plagues of insects with two teeth (probably locusts) were causing 'a great destruction'. Cattle were also experiencing 'a great destruction', and 'many diseases generally reigned all over Ireland'. Perhaps the locusts had been doing what locusts do: eating everything, thus leaving food shortages for bees, cattle and people alike. Or perhaps the locusts introduced an insect disease to Ireland that affected bees already stressed by food limitation. Another 'great mortality' or *duine-badh* of Irish bees, cattle and men was recorded in 992 after a long, severe winter followed by a dry summer and famine.

Fleming also wrote of a 'destruction of bees' that affected the whole of Bavaria in 1035. About four hundred years later, in 1443, poor Ireland suffered another widespread collapse of honey bees. Bee declines were noted in England in the same century. In 1717, there was a 'great mortality of bees' (and carp) in Poland, and the eggs of bees apparently rotted in Saxony beehives over the period of 1780–83, and again in 1796.[10] Fleming's book makes for fascinating reading.

As well as recording animal plagues and diseases for a long time – and not just plagues on land; we also hear from 18th-century writer Thomas Short that 'whales and multitudes of other large monstrous fish were cast on the shore dead'[11] – we have been ascribing and viciously arguing their causes. Comets, eclipses, fireballs, icebergs, lightning storms, demons, parishioners being too nice to cattle, and as many other things as you can imagine have all been thought, at one time or another, to cause dramatic declines and collapses of plant and animal populations.*[12]

Since the publication of Fleming's *Animal Plagues*, we've continued to see and record large-scale losses of honey bees. One of the most famous is the Isle of Wight honey bee epidemic spanning 1905–19. During three different disease epidemics, 90% of the island's honey bees were lost. The causes of this collapse are still debated. Unusual weather likely played a part, as did some dubious and 'disastrous' beekeeping practices intended to remedy bee illness, such as feeding bees formalin or phenol in sugar syrup. A parasitic mite that infected the breathing tubes of bees and a fungal gut parasite have also been blamed, although the influence of both have been concluded as a 'myth' and each called 'scapegoat' by some authors.[13] We've since seen unexplained, large-scale colony losses of honey bees in Canada, Mexico, France, Sweden and Germany.

The most famous, recent widespread colony losses of honeybees was the 'Colony Collapse Disorder', or CCD, first seen on a grand scale in North America in 2006. CCD has a very distinct set of symptoms that differentiate it from other causes of hive mortality: (1) the rapid loss of adult worker bees from affected colonies as evidenced by weak or dead colonies with excess brood populations relative to adult bee populations, with the queen and small number of workers still present; (2) the workers disappear, with a lack of dead

---

* What did people do about plagues in the 14th–18th centuries? There were no pesticides then. Instead, the church was frequently involved in attempts to eliminate pests during the Dark Ages and later. Prayer was just the beginning. Churches and communities established ecclesiastical courts where insect pests were put on trial with legal representation and condemned if found guilty. The offending pests were threatened with excommunication. One of the last trials, in Croatia in 1886, involved a plague of locusts. One of the largest locusts was seized, tried and found guilty (of being a locust?) and put to death. Poor thing. Curses were formally pronounced and the whole species excommunicated from the church. Locusts everywhere must weep, to this day.

worker bees both within and surrounding the affected hives; and (3) the delayed invasion of hive pests (such as wax moths) and robbing or kleptoparasitism from neighbouring honey beehives.[14] In CCD there is also an absence of field-diagnosable bee pathogens that might include bacterial foulbrood diseases and mite infestations or viruses. Parasites such as *Varroa destructor*, a parasitic mite, were introduced in the 1980s and cause hive mortality too, but hives dying due to *Varroa* display very different symptoms, including a declining number of bees, dead bees in or around the hive, frequent robbing by neighbouring hives, and pest invasion.

Nearly a quarter of all US beekeepers suffered from CCD in the winter of 2006–07, with between 750,000 and 1 million of the nation's 2.4 million hives lost. Some beekeepers lost 90% of their hives. Hives that were 'boiling over' with bees one month would have, if anything, just a few young workers and the queen a few weeks later.[15]

The Isle of Wight disease killed 90% of the island's bees between 1905 and 1919. The disease was initially attributed to *Nosema apis*, with remedies at the time including coal tar, hydrogen peroxide, sulphate of quinine, and even pea-flour. *Photo: Library Book Collection / Alamy*

What causes CCD? After the episode in 2006–07, people blamed electromagnetic radiation from cellphone towers. People also blamed genetically modified crops that had a gene inserted for an insecticidal toxin. But we know that this *Bt* toxin is activated and damaging in the gut of some caterpillars, beetles and mosquitoes, and not in honey bees. It's very unlikely that either *Bt* crops or electromagnetic radiation was related to CCD.

Others blamed synthetic pesticides, particularly the neonicotinoid pesticides that have been shown to kill bees and have a range of sub-lethal effects, such as the disruption of the bee's navigational ability. One of the reviews examining pesticide loads in honey bees found reports of 170 different chemicals, including 35 in stored pollen. Although researchers still express concern over the levels of pesticides found in these samples, researchers didn't consider any to be the smoking gun that caused CCD. In many bee colonies experiencing CCD, no neonicotinoids were found.

The upper bee appears healthy; the lower bee is clearly not. As a juvenile, the sick bee has had the parasitic mite *Varroa* feed on it. The bee has a mite on its thorax and a mite to its left. Its deformed wings have left it unable to fly. *Photo: Phil Lester*

Still others hypothesised that poor nutrition and a lack of food sources must be killing these pollinators. Bees need a diverse range of pollen sources to support growth and immunological function. Again, however, nutrition could not be demonstrated as the smoking gun.

About the nearest potential or possible cause that was discovered was the Israeli acute paralysis virus (IAPV). Some researchers hypothesised that a new strain of the virus arrived in the United States in 2005 after the government lifted a ban on live honey bee imports that had been in place since 1922. Experimental infection of hives with this new strain had resulted in bees displaying symptoms similar to CCD.[16] Other reviews downplayed the role of this virus strain, and acknowledged that several 'stress factors' acting alone or in combination weakened hives, allowing opportunistic pathogens to infect and kill colonies. In the end, the broad scientific consensus is that multiple factors contribute to CCD. Pathogens, parasites, pesticides, poor-quality food and climate likely all played roles.

It's been over a decade since North American beekeepers saw widespread CCD. These beekeepers still, however, experience very high rates of overwintering colony loss. Beekeepers in Michigan, Minnesota and Utah report average colony winter losses at 50% or more. I can't imagine how their businesses survive when they lose half their hives every year. Beekeepers in the United States typically experience higher winter loss than 27 of the 29 countries where overwintering mortality is surveyed.[17]

As we've done throughout history, we continue to form strong opinions and conclusions about the causes of these losses. George Fleming's observations still resonate today:

> Men gazed at the phenomena [of epidemics and plagues] with astonishment, and even before they had a just perception of their nature, pronounced their opinions, which, as they were divided into strongly-opposed parties, they defended with all the ardour of zealots.[18]

Beekeepers and scientists discussing the Isle of Wight honey bee epidemic formed just such opposing parties, which they zealously defended. Over the last two decades we have seen just this level of strongly opposed parties and zealots arguing over bee collapses, including on the causes of CCD. More recently, in a discussion of bee declines and collapses, Joachim de Miranda echoes Fleming in noting: 'Each documented decline sparked animated debates across the scientific community discussing the potential causes, generally without a clear-cut resolution.'[19]

Bee declines in New Zealand?

We now have data from 2015–18 on the rates and suspected causes of colony losses within New Zealand.[20] Beekeepers have been extremely helpful by sharing their experiences with honey bee colony losses; in the last survey, 47% of beekeepers responded, which is fantastic – a response rate more than double that for any European country. The US data described above is based on around 13% of their honey bee colonies. In New Zealand, we have data from small backyard hobbyists through to the largest commercial operators.[21] The focus of the survey is winter – a key period for colony losses in honey bees.

What are the results of these surveys of Kiwi bees?

First, I should note that Colony Collapse Disorder has not been observed in New Zealand. I'm frequently questioned about how bad CCD is here. As described above, CCD has a distinct set of symptoms, and an absence of field-diagnosable bee pathogens. Yes, we experience the loss of honey bee colonies in New Zealand, but no, we haven't had widespread colony losses with these specific CCD symptoms.

The average estimated rate of colony loss in New Zealand is about 10%. Colony loss in New Zealand has been estimated at between 8.4–10.2% for the last several years. Some beekeepers, even commercial operators, report losing no hives at all in any given winter. Others report more, with the highest losses exceeding 50% of all hives for commercial or semi-commercial operators. The rates of hive loss are always highest for hobbyist beekeepers.

New Zealand beekeepers were also asked to indicate why their colonies had died. Honey bee colony losses were most frequently attributed to queen

NZ COLONY LOSS SURVEY

SUMMARY 2018

About the Survey

This is an on-line survey of beekeepers that aims to quantify winter colony losses. The survey has been conducted annually since 2015. The questionnaire is based on the international COLOSS survey and has been adapted to include topics of specific interest to New Zealand beekeepers.

Participation

= 25

3,655 beekeepers participated

47% registered beekeepers*

261 beekeepers with >250 colonies

42.5% all registered hives*

*among beekeepers with a valid email address

Estimated Total Colony Loss Rates

8.8% 2016   **9.9%** 2017   **12.8%** 2018

10.7% 2016   **10.5%** 2017   **9.9%** 2018

11.9% 2016   **9.2%** 2017   **8.1%** 2018

10% 2018   **5.3%** 2017   **5.5%** 2016

11.4% 2018   **11.4%** 2017   **7.2%** 2016

10.6% 2018   **9.6%** 2017   **7.4%** 2016

Estimated Total Colony Loss Rates

2018 **10.2%**   2017 **9.7%**

2015 *8.4%   2016 *9.6%

*statistically lower than 2018

Average Colony Loss Rates

9.2% Large commercial   **9.4%** Commercial   **13.5%** Semi-commercial   **33.4%** Non-commercial

Leading Causes of Colony Loss

19.5%
Suspected
varroa

12.1%
Wasps

12.1%
Suspected
starvation

35.5%
Queen problems

Most Common Queen Problems

41%
Drone-laying queens
Old queens 26% more likely

39%
Queen failure
Old queens 48% more likely

Snippets

Beekeepers# report
seeing signs of parasitic
mite syndrome

63%

74%
Beekeepers# report seeing
signs of deformed wing virus
#with greater than 250 colonies

Beekeepers provided any pollination services

37%

At 10.2%, the estimated total colony loss rate in 2018 was substantially lower in New Zealand than in most other countries. The leading suspected cause was queen problems (her disappearance, or failure to lay eggs), followed by *Varroa*, starvation and wasps.

To view the full survey, visit: landcareresearch.co.nz/science/ portfolios/enhancing-policy-effectiveness/bee-health/2018-survey

Image: Manaaki Whenua — Landcare Research

problems. Clearly, as they are responsible for the production of eggs and new workers, queens play a major role in the colony. If they fail, the colony has a good chance of failing too. Nearly 36% of all colony losses were due to queen failure. The queens were typically reported to either stop laying eggs or produce only drones. Both behaviours are frequently observed when queens get old. Beekeepers also reported that young queens occasionally failed, producing only drone brood, or just disappearing, which might be related to mating failure.

The next major problem was suspected *Varroa* infestations and complications associated with the mites' mutualistic relationship with the deformed wing virus. Nearly 20% of all colony deaths were attributed to *Varroa*. It was the cause of a substantial and widespread loss of colonies in New Zealand, occurring over the decade following its discovery in the year 2000. Persistent feral colonies of bees are now rarely observed here, and beekeepers spend much of their resources fending off *Varroa* and its associated viruses.

A further 12% of lost colonies showed signs of starvation. New Zealand now has more managed hives of honey bees than ever before, so there is concern that bees lack food sources. As of June 2019, there were 924,973 registered hives. That's around 3.5 hives for each square kilometre of the country and represents approximately triple the number of hives we had in 2008. John Berry, a long-term beekeeper in the North Island, recently described his frustration with the explosion in hive numbers: 'I go past one of my sites and there are 180 hives dumped across the fence from them. On a farm if you are running a thousand cattle and someone comes in and puts 10,000 cows on the same grass, do you get 10,000 times as much milk? I don't think so.' Beekeepers like John believe they are now in a fight to keep their bees alive – solely because of overstocking.[22] These overstocking rates, potentially leading to starvation, are likely to compound problems associated with parasites, pathogens and pesticides. One hundred and eighty hives dumped next to your apiary will compete with your bees for food, but they also represent a large disease and pest reservoir.

Wasps contribute to another 12% of colony losses. New Zealand evolved with no social bees or social wasps. All of our native bee and wasp species are solitary, and all the social species present in New Zealand today are introduced. Common and German wasps (often called yellowjackets in North America; *Vespula* spp.) raid honey beehives, killing larvae and stealing honey. In the North Island, many beekeepers have reported losing more than 40% of their hives to wasps.

The remaining causes of honey bee colony losses in New Zealand were all reported as relatively minor or infrequent, at <5%. Suspected *Nosema* and other diseases, robbing by other bees, natural disasters, American foulbrood, accidents, thefts or vandalism, Argentine ants, and suspected exposure to toxins such as pesticides were observed at a rate significantly lower than the causes listed above, but also contributed to colony losses. The low rate of loss attributed to pesticides is interesting, given the prominence in mainstream media of pesticides as a cause of bee mortality. I'd note, however, that pesticide effects are often hard to spot. Pesticides may cause a range of sub-lethal effects that weaken colony performance or make hives more susceptible to disease and parasites. Some pesticides have been demonstrated to reduce the foraging efficiency of bees, diminishing food harvest and perhaps making them more susceptible to starvation. Pesticide effects could be hidden in this colony loss survey. I'll talk more about pesticides and their potential effects in chapter 7.

The honey bee industry in New Zealand has many of the same problems as elsewhere in the world, including *Varroa*, American foulbrood, fungal pathogens such as *Nosema*, chalkbrood, and a suite of nasty viruses. But we've got our own challenges too.

Are honey bees the canaries of global insect losses?

The challenges we face in managing honey bees and their habitats may reflect the insect world more broadly. Other insects appear to be suffering from many of the same problems as honey bees.

A recent article reviewing 73 different studies demonstrated that insect biodiversity is threatened worldwide. In their 2019 review, Francisco Sánchez-Bayoa and Kris Wyckhuys found that insect populations have been in 'dramatic rates of decline', which they suggest might 'lead to the extinction of 40% of the world's insect species over the next few decades'. 'Unless we change our ways of producing food,' they write, 'insects as a whole will go down the path of extinction in a few decades.' Alarmingly, they conclude that 'the repercussions this will have for the planet's ecosystems are catastrophic to say the least'. They report some very gloomy statistics, such as those given in a 50-year study in Sweden that repeatedly surveyed large butterflies in a reserve. Of the 269 species initially observed, 59% are no longer found and many of the remainder are in decline.[23] Another study in the review found a 76% decline in flying insect biomass at several German reserves over a 27-year period. Insects in many countries appear to be in big trouble. Populations of pollinators such as bumble bees, which were once widespread and present throughout countries such as the United Kingdom, appear to be only just hanging on. The range of the great yellow bumblebee, *Bombus distinguendus*, is a classic example. Once present throughout the British Isles, its distribution has contracted substantially since the middle of the 20th century. Now it is only known to be present on some islands of the Inner and Outer Hebrides, Orkney, and the northern tip of the coast of mainland Scotland – just a tiny fraction of its previous distribution.[24]

The term 'insectageddon' has been used to describe these results. Headlines have included 'Plummeting insect numbers "threaten collapse of nature"'[25] and 'Insectageddon: New Zealanders have "two weeks of life" after insect apocalypse – expert'.[26] The Sánchez-Bayoa and Wyckhuys review, however, has some issues and critics. It is important to note that they specifically searched for published studies using the search terms [insect*] and [declin*]. Consequently, only studies that found declines in insect species were found. It is an unrepentantly biased perception of the insect population dynamics. It's also worth noting that the vast majority of these very depressing results and studies were conducted in Europe or North America. There were very few studies from the mega-diverse regions around the equator. Countries including New Zealand and Australia might be different. We have few long-term studies of insect populations here in New Zealand other than for species such as invasive wasps, which we have found to be doing just fine.[27] Many other insect pests such as cockroaches, invasive ants, house flies, bedbugs, and mosquitoes seem to me to be annoyingly abundant.

A different story, however, is told by my car windscreen. Driving in summer when I was much, much younger, I remember the sight of our family car windscreen frequently splattered with a wide variety of insects. We had special insect windscreen liquid that was more or less a necessity for summer road trips. It was even worse on my motorbike, because large flying insects felt like bullets smashing into my helmet while riding at (too) high speed. If a large bumblebee flying in the opposite direction hit my helmet, it would snap my head backwards. Nowadays, I need to wash splattered insect guts from

In 2013 New Zealand Post released a series of stamps to celebrate 'the humble honey bee' and raise awareness of the threat of *Varroa*. 'In the years since bees were introduced to New Zealand, beekeeping has developed from a home craft to a progressive industry, and New Zealand is now recognised as one of the world's most advanced beekeeping countries,' NZ Post stated. 'The importance of horticulture and agriculture to New Zealand's economy means that we may be more dependent on pollination from the honey bee than any other nation on Earth.' *Image: NZ Post*

my field of vision much less frequently. That qualitative sentiment is shared by many entomologists in New Zealand and around the world. Habitat destruction, causing a loss of floral or other resources, and the use of the pesticides are cited as candidate mechanisms for our suspected insect losses here too.[28]

Habitat destruction, parasites and pathogens, pesticides, and invasive species influence entire insect communities. But, of all the insects, we probably know honey bees the best. Honey bees can be considered the canary in the coal mine. The colony losses and many of their causes can be seen as indicators of the health of insects in general. An understanding of honey bee health might just have broad biodiversity benefits.

This book

In this book I examine the key challenges facing honey bees. What influences honey bee health? The fantastic work produced by the colony loss surveys from around the world will be my guide. What are the key issues influencing the health and productivity of honey bees? What is being done about these issues?

Chapter 2 reviews honey bee biology and life cycles. Readers will need some understanding of bee biology for the later chapters, including detail on bee development that will influence their susceptibility to *Varroa* mites, or behaviours such as 'robbing' that expose them to pathogens like American foulbrood. Some of you might even be more familiar with honey bees and their life cycle than I am, and so might choose to skip this chapter. I will talk about the 'queen problems' that are consistently identified as the leading cause of colony losses in New Zealand and around the world. The queen problems typically experienced by beekeepers include queen disappearance, slow reproduction, poor patterns of egg-laying and brood (juvenile offspring) development, or queen bees laying eggs that develop only into drones, indicating reproductive failure. A colony is in serious trouble if its queen fails in any regard. But, it's important to remember that every single queen will, eventually, fail. Just like you and me, they will grow old, feeble and unproductive and will eventually die. It's a natural process for a queen to fail. As we will see, the workers will cleverly sense if their queen starts to fail or if she has disappeared. If there are any young larvae in the hive, the workers will feed them a special diet to produce new queens. The old queen will then be unceremoniously 'usurped' by one of these daughter queens that were originally intended to become a worker. The problem of queen failure is substantially worse during the winter months. It means there are no new eggs from which the workers can quickly develop a new egg-laying, worker-producing machine. Queen bees can live for four or more years. But in an attempt to avoid queen failure, beekeepers might replace the queens yearly or sometimes even more often.

Because queen failure is a natural process, I won't spend much of the book discussing this leading cause of colony loss. It is, however, important to note that many of the factors affecting worker honey bees affect queens too. Queens can be susceptible to many viruses and pathogens. Queens, too, are susceptible to pesticides. And a pathogen infection or a sub-lethal dose of pesticides perhaps might not kill a queen, but it may slow her productivity or production of healthy worker bees.

In chapters 3 to 6, I will examine individually the leading causes of honey bee colony collapse or influences on bee health. This is a world of parasitic mites that suck bee blood (well, mostly fat), viruses that shrivel and deform bee wings or turn the adults black and make them shake, and bacteria that dissolve a larval bee into billions of infectious spores that can survive for decades or perhaps even millions of years. In chapter 6 I'll talk about trypanosome parasites and fungal pathogens that give bees dysentery, with a diversion into the world of bumble bees. Then I'll discuss invasive ants and wasps that attack hives, killing the adults, stealing the honey and eating bee larvae. Some of these causes of bee mortality we know reasonably well, though beekeepers and scientists alike are still learning about them.

In chapter 7 I'll review the hefty, emotional and frequently controversial literature on the effects of pesticides on honey bees. Do pesticides and bees go together? Many argue that the answer is 'definitely not'. And not without reason. Pesticides, especially insecticides, are designed to kill insects. Honey bees die every year in every country because of insecticide use. As we saw in the research into colony collapse, honey bees are exposed to and collect many different chemicals. And as mentioned above, the review examining bees dying from colony collapse disorder (CCD) in the US found evidence of 170 different synthetic chemicals, including 35 that were stored in pollen within hives.[29] Bees forage in agricultural and horticultural environments where fungicides, herbicides and pesticides are often frequently used. One of the most hotly debated of these chemical pesticides are the neonicotinoids, which in one study were observed in 75% of global honey samples.[30] These chemicals are probably on your morning toast (though in tiny, tiny, tiny amounts). Neonicotinoids are toxic to insects in very small quantities. Bees will die from neonicotinoid exposure. Perhaps much more commonly, however, honey bees experience sub-lethal effects from these pesticides. Among other effects, they seem to get lost while foraging, have shorter lifespans, and can be more susceptible to parasites and pathogens. A sub-lethal cocktail of neonicotinoids and fungicides might compound the problem.

Many environmentally conscious and caring citizens around the globe have formed their conclusions on neonicotinoids, especially after reading headlines such as 'The evidence is clear: insecticides kill bees. The industry denials look absurd'. This article was in the UK's *Guardian*, which most would agree to be a highly reputable source. The subheading read, 'The largest field trials to date offer irrefutable proof. We need a total ban, now, to halt the sabotaging of our own best interests'.[31] It's no wonder then that we see public protests around the globe to outlaw these pesticides and 'save the bees'.

It is unfortunate, however, that often these articles don't match the science they cite. The *Guardian* article was discussing perhaps the largest field study of neonicotinoids, which was conducted on a grand scale in three European countries.[32] But the results from that study are about the polar opposite of clear. In one of the three countries, neonicotinoid use was associated with an *increased* production of honey bee egg cells, larval and pupal cells within hives, increased worker numbers, and so on. There were no statistically significant effects on hive survival for two of the three countries, with the results in the third country clouded by heavy overwintering mortality due to *Varroa*. Hives in each of the three countries displayed quite different results in this study.

My goal for chapter 7 is to critically examine the science associated with pesticides and honey bees. I want to do this review in an unbiased and open-minded way, and would urge readers to do the same. While most of us would agree that it would be wonderful to have no synthetic chemicals anywhere near our bees – or any of the world's biodiversity – my goal here is to step back from emotion and preconceived ideas and to critically examine pesticides, including neonicotinoids.

In chapter 8 I'll examine other problems for honey bee health, including some species that are usually a minor bother for beekeepers and others that are peering over the horizon as major threats. These include small hive beetles, moths, and hornets that hover in front of hives to catch and kill tired foraging workers as they return to their sisters.

Beekeepers and fancy-dressed supporters take to Parliament Square, London, in April 2013 to call for a European ban on the use of neonicotinoids.
Photo: Patricia Phillips / Alamy

The final chapter of the book will look to the future of bees and beekeeping. I'll discuss starvation, which is considered a significant cause of honey bee mortality according to the colony loss surveys in New Zealand and around the world but is something more of a management issue. What priorities and challenges should we worry about now in order to see better bee health in the future? What should be our management and research emphasis? Working now to improve honey bee health for later will benefit the honey bee industry, our horticultural and agricultural industries, and biodiversity at large.

The goal to 'save the bees' is admirable, but how can we do more?

Thirty billion bees are moved each year into California for the pollination of almonds, in 'the largest managed pollination event in the world'. The transport and monocultural diet of almonds is stressful on the bees. *Photo: Della Huff / Alamy*

1. THE LIFE HISTORY OF BEES

Democratic decisions and the occasional coup d'état

If a queen honey bee is ever abandoned by her workers, she will die. You won't see a new queen honey bee leave her hive to establish a new colony on her own. You won't see her forage for pollen or nurture her eggs to adult bees, or defend the hive from robbers who want to steal honey or eat the larvae. The queen is always reliant on her daughters or sister workers for these tasks. These workers are essential to the success of the queen, just as the queen is essential to the workers.

The hive is a 'superorganism'. A superorganism is defined as a colony of social individuals who, by self-organisation, effective communication and through effective division of labour, assemble into a highly connected group or community that functions as if it were a single organism. That definition nicely describes a beehive. One of the pioneers of modern beekeeping, German beekeeper Johannes Mehring (1815–78), went as far as giving these honey bee superorganisms the status of honorary vertebrates. The queen and the male drones represent the female and male reproductive organs, he argued, while the digestion and body maintenance are roles for the workers.

Poor Johannes was ridiculed for this analogy. But more recent authors have gone further, giving these insects not just vertebrate but honorary mammalian status.[1] The reasoning goes that bees can produce nourishment ('milk') for their offspring from special glands. This milk equivalent is the royal jelly produced by nurse workers and given to all larvae in a colony. Looking for further similarities to mammalian biology, we could say that the queen represents a social uterus for the superorganism; bees have an immense capacity for learning, with a cognitive ability that exceeds that of many vertebrates; and the young larvae have a high body temperature, similar to that of mammals, at around 35°C. Honey bees also have a relatively low rate of reproduction, which we also see in many mammals.

Although worker bees are produced continually in warm weather, a new bee colony, or a new superorganism hive, is typically produced only once per year.

Bees as honorary mammals seems a bit of a stretch to me, but I can see why the analogy has been drawn. There is no question that bees are extremely successful and have a fascinating biology. That biology can include a princess attacking and viciously killing her sisters. It can include 'sterile' workers secretly laying viable eggs, and male penises exploding during sex with an audible pop. Bees' ability to communicate the location and quality of food to their sisters is astonishing, as is their ability to learn.

To start this book, I'll describe the wonderfully diverse life within a honey bee colony. An understanding of the inner workings of the hive or superorganism will help in understanding the issues that bees face, which we'll survey in later chapters.

Beekeeper Frank Lindsay working his beehives near Wellington. *Photo: Phil Lester*

The young worker has stay-at-home roles

There are three different life forms of the honey bee: the worker, the male drone, and the queen. The most numerous form in the hive is the worker. A large hive in summer might contain 50,000 workers. These workers are all female and are produced in spring, summer and autumn. A worker has a remarkably short developmental period: approximately three days as an egg, eight days as a white wormlike larva continually demanding food, and nine days as a pupa, before chewing its way out of its cells as an adult. These adult bees experience a 1500-fold increase in weight over those three weeks.

The newly emerged worker bees might live for just a month in summer as adults, or as 'winter bees' they might persist for four to six months over the coldest season of the year.

During their adult life, these workers take on a range of roles. Typically, the young adults do tasks within the hive, including brood care and food processing, and in their later days move to foraging outside their hive. Foraging is a dangerous task and is associated with substantial increase in mortality. Honey bees don't do a lot of foraging in winter, which is one of the reasons over-wintering workers live longer. Bees won't forage at temperatures below 13°C, and the colony won't achieve full foraging mode until a temperature of 19°C is reached. In summer, each worker bee might visit 50–100 flowers on a foraging trip. The workers need to visit around three million flowers in order to produce a kilogram of honey.

Honey bees exhibit a 'sort of' age polytheism: usually they do different tasks as they get older. That 'sort of' is because it isn't a strict rule, and bees may do tasks in the hive according to colony needs.

First, newly emerged workers (2–15 days old) are typically on cleaning duties and nursing the larvae. Nurse bees are effectively the social stomach that digests pollen into vital protein for the entire colony. These young adult bees use enzymes to break down the pollen, then they secrete a jellylike substance that is fed to larvae, adults and the queen. You are probably aware of what happens to someone's digestive system if they are lactose intolerant and drink milk. A similar thing happens to older, forager bees: they lose the ability to digest proteins in pollen. Any pollen they eat shoots through their gut largely undigested. To get a protein hit, the older workers have to beg for jelly from the young nurse bees.

Tending the young is a strenuous task. Until he was two years old, my youngest son refused to sleep for more than two consecutive hours. It was miserable. I'd get up at 4am with him every day so as to let my wife have a few hours of decent sleep. I know I was grumpy as hell for those two years. My work almost certainly wasn't of the quality it should have been, and I struggled to communicate even in short sentences. Bees experience similar issues. A recent study with bumble bees demonstrated that when there are large numbers of larvae to be fed, nurses and other bees get substantially less sleep.[2] (All bees sleep, and there is evidence that they might even dream, because deep-sleep phases in honey bees seem to prompt memory consolidation, just as they do for you and me.)[3] Sleep-deprived bees are sloppy in their communication, inaccurate in their ability to point the sisters to food, with waggle dances becoming much less precise.[4] And then they get lost on their way home to the hive.

As the bees age, their hypopharyngeal glands (their equivalent of mammary glands) stop producing jelly, and these bees move on to a new role.

Ever wondered where the wax comes from in a beehive? At the grand old age of 14 days after adult emergence, or thereabouts, the worker begins to secrete wax from her wax glands. The two-week-old adult honey bee starts the process by gorging herself with honey and literally hanging out in a cluster of bees. About a day later, small flakes of wax are produced on her abdomen. The wax flakes are chewed, mixed with secretions from other glands, and then passed around the hive for building combs, repairing cells, or capping the cells of larvae that are turning into pupae. Producing wax is costly, demanding a considerable amount of energy. Scientists have estimated that it takes 8–15 kilograms of honey to produce just 1 kilogram of wax.

Two-week-old adult bees do a range of other tasks. 'Receiver bees' unload nectar from foragers and process it into honey by removing a substantial amount of water. Honey is around 15–18% water. Because of its low moisture content, no bacteria can live in it, and the high sugar concentration kills microorganisms by plasmolysis (in other words, the honey sucks all the water

1. A queen bee lays an egg in the brood cell.

2. The egg develops and hatches.

3. Workers feed the larva, which goes through five growth stages.

4. Workers cap the larval cell. The larva finishes developing, and pupates.

5. An adult bee emerges.

The honey bee, from larva to adult.
Illustration: Bayer Scientific

from the bacterial cell and bursts it). To reduce the water concentration of nectar and refine it into honey, bees repeatedly drink and regurgitate the liquid. During this process, they add a range of enzymes that have roles like changing the sugars from sucrose into equal parts glucose and fructose. An enzyme called glucose oxidase is then added, which breaks down the glucose and stabilises the honey's acidity. The resulting liquid is called ripened nectar and is smeared over wax cell combs. It is then fanned by the worker bee wings for further water evaporation.

The teaspoon of honey that you spread on your toast represents a lifetime of foraging by a dozen bees. You might also think about the bee digestive systems in which that yummy liquid has been: honey has been vomited and sucked up many times, by many bees, before you put it in your mouth.

Older workers typically do the foraging

The worker becomes a forager at about three weeks after emergence as an adult bee. There are four major needs for the colony: nectar (a carbohydrate or sugar source), pollen (the protein for the colony), propolis (for hive construction and repair) and water.

Nectar is composed of sugars and waters and is produced by plants for attracting insects, which are typically pollinators. Bees will use their straw-like proboscis or modified mouthparts to suck up nectar. The nectar is stored in the first part of the bee's stomach. The foraging bee will then return to the hive after visiting perhaps hundreds of flowers on a single trip to regurgitate the nectar to other bees. The flowers from which she collects are typically all the same species, resulting in individual bees showing a high degree of 'crop consistency'. When a bee has found a plant species she likes, she will stick to foraging on that species. If she has found a good nectar source, she might communicate its location to other foragers with a waggle dance. First, she'll feed her sisters some of the nectar she has discovered, then she'll do a figure-eight dance that shares information about the direction and distance to resources such as patches of flowers. There are other dances and communication patterns in the hive, including the 'round dance', to indicate nearby food sources, and a 'grooming dance', which encourages a bee's sisters to clean.[5]

Bees are discerning creatures and clearly respond to the amount and quality of nectar produced by flowers. From experiments in which scientists alter the sugar concentrations of water at bee feeders, we know that foraging workers can determine the amount of sugar in liquid and that sweeter nectar sources attract more bees. But is the opposite true, and do plants respond to bees? Can plants hear? There is surprising recent evidence that, yes, plants seem to be able to sense and respond to airborne sounds.

A group led by Lilach Hadany at Tel Aviv University in Israel have recently shown that primrose flowers quickly produce sweeter nectar within three minutes of being exposed to the sound of a flying bee or even to synthetic sound signals at similar frequencies.[6] The theory is that the plant responds to their presence by increasing the chances of pollination. The flowers are effectively the plant's ears.

They vibrate in response to typical sounds of bees but not to higher or lower frequency sounds. That plants might hear seems astounding, but perhaps it shouldn't be that surprising. We've known for a long time that some plants have flowers that release pollen only upon vibration by some specialised pollinators. This process is called buzz pollination. Species like the bumble bee will grab a flower such as a tomato flower, detach their wing muscles and vibrate their bodies, thereby shaking off the pollen. Honey bees don't do buzz pollination. So, if you want tomatoes and many other crops, you need bumble bees.

Another group at Tel Aviv University have data that suggests some plants make high-pitched noises that might attract pollinators some distance away, although this hypothesis has been met with a degree of scepticism.[7]

Bees are herbivorous and require pollen for their protein, vitamins, fats, minerals and trace elements. Protein is especially important for the growth and development of larvae. A large, strong hive might collect 15–40 kilograms of pollen a year. Unlike nectar, pollen isn't stored in the digestive system for the trip back to the hive. Instead, honey bees have special 'pollen baskets' on their hind legs. The first and second pair of bee legs gather the pollen that is caught on the very hairy bee body and, using a pollen comb, transfer it to the baskets. Hairs on the hind legs keep the pollen in place as the bee flies around. The total weight of pollen collected on a foraging trip, in which the bee might visit hundreds of flowers, might weigh 25 milligrams. Back at the hive, other workers unload the pollen. During this unloading process, the pollen is mixed with salivary secretions and probably some nectar. It is stored in open cells on the honeycomb at the interface between the brood and the stored honey. This mixture of pollen and secretions is called bee bread and is the primary protein source for the larvae and workers. Depending on the pollen source, bee bread might contain 60% protein, exactly what is needed to quickly raise more baby bees.

The pollen that foraging bees bring back typically has a high water content and so should be ideal for mould growth, especially in the nice warm environment of a beehive. During the pollen handling, however, the digestive juices of the bees are added to the pollen mix. These digestive juices inoculate the pollen with lactic acid bacteria. These bacteria proliferate best in anaerobic conditions (when oxygen is present only at very low concentrations). Lactic acid bacteria are also restricted to environments in which sugars are present. In the storage cell of the beehive, the pollen pellet is typically sealed from the air with a layer of honey.

Historically, it was thought that the bacteria dissolve the pollen sheath and metabolise the sugars in the pollen, producing lactic acid. More recent research, however, suggests that the bacteria's role is solely in the storage of the pollen, not pre-digestion.[8] The bacteria lower the pH of the pollen/saliva/nectar/honey mixture from 4.8 to around 4.1, which is a below the threshold for the growth of many pathogenic bacterial or fungal species. The final pollen-based product is bee bread, which can be stored for long periods.

The protein that bees get from pollen is clearly important for their development. If honey bee larvae are deprived of pollen, they grow up to be poorly performing adults. Worker bees without adequate access to pollen forage for fewer days, are more likely to die after only a single day of foraging, are less likely to waggle dance than their unstressed counterparts, and, if they do dance, frequently give imprecise information on the location of the food.[9]

Head down, a worker packs pollen into a cell within the hive. The pollen is mixed with salivary secretions and probably a little nectar. It is stored anaerobically, without oxygen. Bacteria from the bees convert the pollen mixture into bee bread. *Photo: Phil Lester*

Propolis is the third category of material that bees make from the raw materials they forage. It's sometimes referred to as 'bee glue', because it is a very resinous mixture. Propolis is much harder than beeswax, and the process of making it typically starts with collecting sap from trees. The varying colours and shades of propolis reflect the different tree species from which the resin has been collected. In the hive, again, bees convert the resin to propolis by mixing it with wax and saliva. It is a complex mixture of up to 300 different chemical components.[10] A primary use of propolis is to seal holes or gaps around the hive in order to stabilise the structure and provide insulation, and also to defend the hive against predators like wasps and hornets that would like to raid the honey and eat the brood.

Larger predators can invade the hive too, like a mouse seeking food or warmth. Should this mouse be discovered, it would be repeatedly stung and hopefully driven off, but if it cannot escape in time the rodent might die within the hive. Propolis would then be used to seal and mummify the mouse carcass in order to stop it culturing disease within the hive. The propolis does much more than seal: it seems to have substantial antimicrobial properties and pesticidal activities.[11] There is good evidence that it can inhibit the growth of diseases within the hive, including the bacterial pathogen causing American foulbrood. Propolis even seems to have miticidal activity, in that it can kill the parasitic mite *Varroa*. One study showed that bees in colonies experimentally treated with propolis showed a reduced expression of immune genes.

Absolutely amazing. That's how I'd describe wax, propolis, the storage and manipulation of pollen that is converted into bee bread, and nectar that is converted into honey. All of these conversions require the evolution of specialised enzymes and bee-produced chemicals.

We humans have been infatuated with all things bee for millennia. This fascination has extended to the possible virtues of bee bread for human health. It's been suggested that consuming bee bread can strengthen your immune system, support the treatment of pharmaceuticals, improve your concentration and memory, regulate your digestive system functioning, regulate cholesterol levels and reduce total lipid content in the blood, have anti-ageing and anti-anaemic effects, regenerate all the cells of your body, and benefit the purification of your liver.[12] If all this were true, bee bread would be about the most valuable substance on Earth. Similarly, some authors have suggested that propolis can cure cancer (well, at least it has anti-tumour activity in breast cancer cells), has antibacterial capacities for oral, gastrointestinal and vaginal infections, can help heal wounds, and a whole lot more.[13] I'm sure there is some evidence for many of these claims. But much more work is needed to confirm such benefits before I'll be sprinkling bee bread onto my cornflakes in the morning, or relying on propolis if I ever develop cancer.

A mouse skull covered in propolis. Having decided to investigate a beehive, the mouse died in the hive, perhaps after being stung. Honey bees can't move a large animal from the hive, so instead they encase the carcass in propolis. *Photo: Marla Spivak*

Guard bees and their stings

Because their stinging ability and venom production is limited, young adult worker bees would make very poor defenders and guards of the hive. As adult bees age, they develop these defensive weapons and are more likely to become guards of the hive entrance, keeping a watchful nose (via chemical receptors on the antennae) and eye on anyone attempting to enter the hive, such as honey bee workers from other hives. Other potential threats include wasps, ants, moths, mice and people. Guards have a typical posture, standing at the entrance on their middle and hind legs, with their forelegs raised and bristled to inspect arrivals.

One recent review suggests that 10–15% of honey bees take on guard roles when they are two to three weeks old.[14] That proportion of bees on guard duty varies a lot, however, due to factors such as nectar flow and the number of potential threats. More wasps buzzing around the hive mean more bees will be on guard duty, and these guards will be aggressive and on high alert. Guards can make mistakes, though. They appear to use odours or a form of 'recognition template' to differentiate nestmates from intruders. Chemicals called hydrocarbons form the recognition template, which is effectively the average odour representing their colony. If the smell is too dissimilar, rejection occurs. Sometimes colonies smell similar, and foreign workers are accepted by the guards. About 30% of drifting bees are allowed to stay. These workers may have been planning a robbery, but they may also have just gotten lost and innocently drifted into a new and nearby hive. The drifted bee might then spend the rest of its life working in the new hive, or drift back to its birth hive.

Guards have different patterns of gene expression from foraging bees. As a result, their brains work differently. Some researchers have suggested that guards and comparatively aggressive bees are often infected with a virus. In some honey bee populations, there seems to be a close relationship between a pathogen called the Kakugo virus present in the brain and aggressive behaviour such as that seen in bee guards.[15] Higher levels of aggression appear to be driven by a faster and greater response to this alarm pheromone. The European honey bee may attack intruders that venture within a few metres of the nest, then aggressively pursue these intruders to around 50 metres from the colony. By comparison, the Africanised honey bee (the subspecies *Apis mellifera scutellate*) might initiate an attack on an intruder at a distance of more than 100 metres from the nest and then pursue the fleeing intruder for several kilometres.[16] Upon perceiving a threat, the bees will exude alarm pheromone from their stingers, which is reinforced by when actual stinging takes place. So, that sting you receive marks you as a target for more stings. Different strains of bees appear to be more aggressive than others, with the Italian honey bee strain (*Apis mellifera ligustica*) renowned for its gentler, less aggressive nature.

The guards and other bees will quickly sacrifice their lives for the hive by stinging potential intruders. When they attack mammals such as us, their barbed stings are ripped from their bodies along with many of their abdominal organs. That action is lethal for the bee.

How much does a bee sting hurt? The entomologist Justin Schmidt has spent a world-famous career classifying the painfulness of hymenoptera stings from a

A guard bee stands on her hind legs with forelegs raised to inspect a new arrival. Guard bees spend most of their time inspecting foragers who arrive at the nest, but they're also ready to defend against intruders like ants (which they try to fan away) and mammals (which they sting). *Photo: Phil Lester*

wide range of species, including honey bees. He has been stung by more than 83 different species from all corners of the globe, in order to inform me and you exactly how much it hurts. He has been named the 'The King of Sting' and the 'Connoisseur of Pain'. One of his major contributions is the Schmidt Sting Pain Index, a four-point scale for hymenoptera stings. The sting of a honey bee, which rates only 2 on that scale, is described as 'a flaming match head [that] lands on your arm and is quenched first with lye and then sulfuric acid'. That sounds bad enough. But it can get worse, especially if one day you are riding your bike with an open mouth into which a honey bee flies. The result of that sting on your tongue is 'immediate, noisome, visceral, debilitating. For 10 minutes life is not worth living'.[17]

Another entomologist, Michael Smith, 'rated the painfulness of honey bee stings over 25 body locations in one subject (the author)'. He was duly stung on various painful locations including fingers, toes, hands, feet, legs, lips, nostrils, nipples, scrotum and penis. Surprisingly, a sting to the nostril took the cake. 'A honey bee sting to the nostril is a whole-body experience. Your eyes tear up, your nose is spewing mucus, you're sneezing, and all you want to do is get that stinger out.' Smith swears that 'if you are forced to choose between a sting to the nose or to the penis, you're going to want more stings to the penis'.[18] My personal preference is none of the above.

Different people react differently to being stung. Jess, a graduate student in my lab, was stung on her eyebrow recently and it looked as if she had been in a car accident or on the losing side of a boxing match. Our lab group was always excited when Jess worked the bees. You never knew what the outcome would look like.

My reaction to bee stings has changed over time. Many decades ago when I was young, after a bee sting to my toe my whole leg would swell up. I'd cry (it hurt). Over time, my reaction and range of emotions towards bee stings have mellowed. But my worst stinging event ever was more recently. I had a quick job to do – grab a few frames of bees from a rooftop hive at the university to put into our observational hive for a public display. Our bees are normally docile and friendly, so I put on a protective suit, but not boots. Alas, that day our bees were not so friendly: perhaps some wasps had been flying around the hive and the bees were on high alert. And I found that the job was not so quick because there were some issues with the hive that I had to fix while I was there. The combination of these events meant that what seemed like a legion of angry bees found their way up my pants leg and into my bee suit. I received about 20 stings and swore a lot. Jess was there and thought it all hilarious.

That level of stinging might not have been quite so bad if earlier in the morning I hadn't had a series of vaccinations for a trip to South America. The combination of stings and vaccinations seemed to overload my immune system. I probably had what is called a cytokine storm or cytokine release syndrome. For the next two days my immune system went into overload with a fever, elevated heartbeat, and a range of other symptoms I'd rather not describe. It wasn't fun. My next sting was about a year later while taking photographs of *Varroa* and American foulbrood for this book. This time, boil-like hives appeared over much of my body. I look forward, with some trepidation, to my next sting and I carry an EpiPen now.

Honey bees are almost certainly the animal that kills the most people each year in countries like New Zealand. In Australia, more hospital admissions occur from bee and wasp stings than from 'scarier' animals like snakes and jellyfish.[19]

Winter worker bees

Another group of bees that have a distinct role in the hive are the winter bees. These are the bees that live for four to six months. Winter bee workers are different from workers born in summer, with some biologists believing they are even a different 'caste' or life form. They fit the definition of castes, as they are physically distinct and have a specialised role in the colony. Basically, winter bees are fatter, slower and don't get out much. They have larger-than-normal organs called fat-bodies in their heads and abdomens, which are used to break down not only fats but also carbohydrates (like in honey), proteins, and other molecules. They act as protein stores and reserves for the colony.

Winter bee production seems to depend on diet. Pollen is the currency of a beehive: energy is obtained primarily from nectar that is turned into honey, but pollen provides everything else. Winter bees seem to be produced by a reduction of pollen in the larvae experienced in autumn.[20] This long-lived caste of winter bees is sometimes referred to as diutinus – a pretentious word for 'long-lived'. (Feel free to use 'diutinus' if you are playing Scrabble, but 'winter bees' is better if you want to be understood.)

Some workers produce viable eggs

There is a preconception that all worker bees – perhaps around 50,000 in one hive – are sterile and unable to reproduce. But no – the workers do have ovaries and can produce viable eggs. Life in a beehive is dark and typically controlled by sound, touch or pheromones. Honey bees have at least 15 different glands that produce a wide array of chemical pheromones or intra-species communication chemicals, including the 'queen mandibular pheromone'. This and other pheromones have a variety of roles, including to inhibit ovarian development and egg-laying in worker bees. For workers, the presence of these pheromones is a critical cue that indicates a queen is present in the hive. Workers tending and cleaning the queen become covered in this pheromone, which they then spread throughout the colony as they move around.

Even in the presence of a viable queen and her fragrant perfume, perhaps one in 10,000 honey bees may still become reproductive and lay eggs. But in the absence of a queen and this pheromone the ovaries of around 10% of workers in the hive will develop ovaries and produce eggs. Though, in a weird kink of insect sex, all of the eggs will be males. The workers haven't mated and so have no sperm to fertilise eggs. Only fertilised eggs become females (the haplodiploidy mating system). Somehow, however, other workers often recognise worker-laid eggs and will kill and eat as many of them as they can. Effectively they police the eggs, and will smell out and beat up their reproductively active sisters.

Males do nothing around the house and live for just one penis-popping purpose

This morning I read an article in the *New York Times* that suggests that men in Japan do almost nothing around the house. The article cites an economics professor at a university in Tokyo, who found that women who work in excess of 49 hours in a week also have the privilege of doing an additional 25 hours of housework. Their working husbands spend an average of less than five hours on domestic duties a week.[21] Instead of washing the dishes, many Japanese men are culturally obliged to go out drinking with their work buddies each night. From my male-centric perspective, Japan sounds immensely appealing.

Drones, or male bees, have it even easier than that, contributing much less than five hours per week. They do absolutely nothing in the hive. They don't forage, won't participate in hive defence or feeding the young, and have no wax glands or sting. Their sole function is sex.

Because drones are produced from unfertilised eggs, they have half as many chromosomes as the females (16 compared with 32). They develop, look and behave very differently. Drones take the longest amount of time to develop in the hive: around 24 days to develop from egg to adult, compared with just 16 days for a queen and 21 days for a worker bee.

Drones are produced in hives primarily in spring and summer, with peaks in production when colonies are swarming. There might be a few hundred or thousand drones produced in a colony over any one season. The number produced depends both on the worker colony size, available space, and the amount of food in the hive. Some evidence suggests that drones are costly to produce. Without the waxy drone comb, honey bees may produce more honey,[22] although other researchers have found evidence that they have little impact on hive productivity.[23] It seems logical to me that drones would be costly to the colony. They are fed honey and the pollen-rich bee bread as larvae, then as adults these males help themselves to as much honey as they want. As they don't do anything to help with the production of the food they consume, how could they not be costly?

After emerging as an adult, the drones begin to fly after around eight days and become sexually capable after 12 days. Then the search for a date begins. They have a nice sleep over the evening and through the morning, they have breakfast and lunch uninterrupted by any paternal duties of childcare, then in the afternoons they take flight for an hour in search of virgin bees. These lust-filled and hopeful males often fly to 'drone congregation areas'. A drone congregation area might be 30 metres off the ground and have a diameter of 100 metres, sometimes containing thousands of honey bee males anxiously looking for anything vaguely resembling a virgin queen. These are the nightclubs of the bee world.

We don't know why drones decide upon any one place as a drone congregation area. Some people think that it is due to the earth's magnetic field, or specific landscape features. All we really know is that drones will fly up to 3.75 kilometres to find a congregation area, in which they will compete with one another to fatally mate with a virgin queen. There might be drones

Research scientist Eloise Hinson catching drone bees for research into bee genetics. To catch drones you lift a net into a drone congregation area with a balloon. A virgin queen can be placed inside the net, or a lure containing (E)-9-oxo-2-decenoic acid, which is the queen substance of the honey bee.
Photo: Ben Oldroyd

present from over 200 different hives.[24] Their powerful flight muscles and stocky bodies have been evolutionarily tuned for this precise moment. Throwing a stone into this hormone-riddled pack of eager males will create a speeding rush of optimistic excitement. An actual queen entering the area, rather than a rock, attracts male interest not only because she moves, but also because she is reeking of her special perfume. Queens secrete a range of chemicals, including queen pheromone, which attracts males from hundreds of metres away. The chase begins. A 'drone comet' forms in hot pursuit, with the males typically flying beneath the queen in a pack. Their big eyes on the tops of their heads keep her in their vision by silhouetting her against the sky. She turns and races, leading the hopeful pack on.

The first successful male captures the queen while flying, grabbing her abdomen with all six legs. His abdomen curls under the queen, who opens her sting chamber. Sex in honey bees lasts approximately three seconds, which must rank among the shortest time for any species anywhere. Once an opening is sensed by the drone, he will evert his penis instantaneously and ejaculate about six million sperm. With an audible pop, his genitals then explode as he separates from the queen and falls to the earth, paralysed. He will die within minutes. That's it for the successful drone, but the queen will mate again with the same routine, except the next male must first remove the mucus plug and bits of exploded penis left from the previous guy.

Honey bee queens have been described as displaying 'extreme hyper-polyandry'. That means that any individual queen will mate with *many* different males. She might have over 60 different partners.[25] This unusually generous level of promiscuity might be a response to environmental and bee management conditions, as there is evidence that a high level of genetic diversity promotes colony functioning and strengthens the hive's ability to overcome disease and other stresses. Perhaps one male lineage is susceptible to a particular disease, but another might carry immune genes that show resistance. So, some eggs and larvae fertilised by sperm from one male might die, but the eggs fertilised by sperm from the disease-resistant male will live. Experiments have demonstrated that colonies with higher genetic diversity, from multiple mating, typically have less disease, are better able to control the hive temperature, have greater foraging and worker recruitment abilities over larger areas in the environment, have more workers that store more food, and are more likely to survive the winter.[26] All this evidence suggests that more sex and higher levels of genetic diversity equals a better, healthier hive.

Once her spermatheca (sperm sac) is full, she will never mate again. Those sperm from all of her partners will live and can be nourished inside her spermatheca for years.

Unsuccessful drones must lament that they were sufficiently unfortunate to live. They will return to their hive to relax, refuel, and try their luck another day. These sadly alive males with unexploded genitals are tolerated by the workers, at least while there is a chance that they might win the drone-comet race to catch a virgin queen another day. The workers will, however, eventually lose their patience. Consistently unlucky drones who live to see the end of the mating season are aggressively evicted from the hive by their worker sisters. Cut off from their larders of food and unable to forage, or do anything for themselves, they quickly die.

Gynandromorph bees are reproductive oddities. This gynandromorph has a large male-type eye but many female characteristics. *Photo: Sarah Aamidor*

Some weird things can happen with honey bees and reproduction. About the weirdest is gynandromorphy. Gynandromorphs are individuals that have both male and female characteristics, resulting from a mix of male and female tissue. You'll occasionally see vertebrate gynandromorphs in nature, like a gynandromorph cardinal that was spotted in Pennsylvania in 2018. The bird was bilaterally divided, with the bright red of the male cardinal on its left, and the brown of the female on its right.[27] In honey bees this effect can be quite common, resulting from more than one sperm simultaneously entering the egg cell. A recent study even found a gynandromorph honey bee that had three fathers and a single mother.[28]

The resulting gynandromorph bees can only be described as a frankenbees: they look like something a mad scientist has assembled from many different and unwilling donor bees. Even more bizarre is the recent discovery of a honey bee with two fathers and no mother. As I've discussed, unfertilised eggs laid by the queen typically result in males. Those males have one set of chromosomes, while females result from two sets after the eggs are fertilised by a drone. In the case of the honey bee with two fathers and no mother, the diploid individual seems to have resulted from two sperm fusing together, resulting in a normal-ish-looking female bee.[29] This fusion happened within the female reproductive system, but she contributed no genes or chromosomes to the resulting egg.

A princess is born! And now there is trouble in the hive

We tend to think of a honey bee queen as the leader or chief decision-maker in a hive. That description is not at all accurate. In reality, the queen is an egg-laying machine. After her initial mating flight and return to the hive, about all she does is lay eggs and eat in order to produce more eggs. She will lay up to 1500 eggs each day in spring and summer. Her workers are the true decision-makers, closely monitoring and regulating nearly everything in their environment. They regulate the queen's reproductive output and play a major role in deciding on a new queen.

When the queen lays an egg, she decides whether that egg will be fertilised or not, and, if not, that egg will develop into a male. Usually she'll produce a fertilised egg by releasing just a few of the sperm acquired during her flights through drone congregation areas.

A fertilised egg has the potential to develop into either a worker bee or a new queen. Its fate is dependent on the food provided by the young bee's sisters. If the young bee is given a mixture of bee bread derived from pollen, nectar and royal jelly, a worker will develop. If instead the young bee is solely given the royal jelly produced by the nurse workers, a queen will emerge from the cell. A royal jelly diet raises the level of juvenile hormone: by the larva's third day of life, consistently high levels of this hormone mean that its caste and fate are sealed. Queens are raised in larger cells called queen cups, because queen bees are a little larger than the workers. They are produced to emerge, fly and mate in spring and early summer.

Insects such as wasps produce queens that mate and overwinter by themselves. Honey bees are very different. Their primary reproductive mode is swarming, wherein the mother queen and approximately three-quarters of her daughter workers in a hive will leave the colony immediately prior to new, adult virgin queens emerging from specially built cells. I'll discuss the production of these virgin queens before coming back to this swarming behaviour.

New virgin queens are produced in the queen cups. Normally, several virgin princesses or potential new queens are produced. Only one of these virgin princesses will emerge victorious as the new queen, with the others largely as an insurance policy for the hive. The workers monitor the developing queens and sometimes kill up to 50% of them, apparently in an attempt to weed out the weaklings so they are left with only the high-quality individuals.

The very first task of the newly emerged adult princess is to hunt down and viciously attack her sisters. They detect each other in the dark hive by chemical pheromones and by beating their wings to make a 'tooting' or 'piping' sound. Princesses still in their cells, not yet emerged, may respond with a similar sound called 'quacking'. Most people don't know that bees quack.

These several virgin princesses or potential new queens are aggressive. They will chew holes to enter developing queen cells, stinging and killing their younger sisters. If another emerged virgin queen is detected, a fight begins and may last for anything between five seconds and 15 minutes. The two large bees grapple with each other while attempting to use their stings. The fighting is intense and often the bees deploy a very dirty trick, attempting to discharge their faeces onto each other. A virgin princess bee covered in faeces will be immobilised by the worker force and less likely to win the battle, partly because the other queen can sting her while she is immobilised.[30]

The losing bee is stung and paralysed. She falls to the floor of the hive where she may receive more stings from her victorious sister. Fifteen minutes later, it is all over for the fallen virgin.

You might think that the workers are all bystanders, passively watching and waiting for the outcome of these gladiatorial battles between their sisters who would contend to be queen. But no – the workers often substantially influence the outcome of these interactions. We don't fully understand why the workers might favour one of these princesses over another, but given it is dark in the hive their decisions must be related to the quality of the smell they produce and/ or the piping sounds they make. The workers might re-seal the cell cap of an

A queen bee surrounded by a retinue of workers. Queen retinue pheromones help her to communicate with the workers in the dark beehive.
Photo: Phil Lester

emerging queen cell, stopping her from exiting her cell and fighting other queens. They might swarm over and protect developing bees still in their queen cells. They might protect some virgin princesses while chasing and harassing others, ultimately determining the victor in these battles. Perhaps these behaviours help ensure that the individual who succeeds is the strongest virgin queen in the hive. The victor and future queen arises within 24–48 hours of the first virgin princess emerging. The carcasses of her defeated sisters are removed and discarded from the hive by the workers.

Meanwhile, the mother queen and a swarm of around three-quarters of the bee population have left the hive. Initially they fly just a few metres from their old residence, where they congregate in a bee beard.

In the weeks prior to the swarm, a population explosion occurs in the hive, resulting in large numbers of workers. A few days prior to swarming, the worker bees reduce the amount they are feeding their 'old' queen. The workers

Artificial insemination of a queen bee at Wurzburg University, Germany. The genital opening has been teased apart and a syringe is used to inject semen (white) from selected drones. Artificial insemination allows researchers and breeders to control the genetics of a colony, the members of which are all offspring of one queen. Desirable characteristics may be selected in this way.
Photo: Simon Stone / Alamy

also appear to be mildly hostile to their queen around this time, grasping her and lightly biting, shaking and pushing her around. These sessions of rough handling are almost continuous near the swarming period, forcing the queen to keep walking and moving around the hive. This enforced exercise regime and calorie restriction does wonders for her waistline, with the queen experiencing a dramatic reduction in abdomen size. Her ovaries shrink and now she can easily fly.

Immediately before swarming, the worker bees gorge themselves with honey from the old hive. Then they initiate the swarm by starting a special type of waggle dance, or what some describe as a 'whir dance'. They run forcibly between other bees in a zigzag fashion, vibrating their bodies and producing an audible whir. Other bees join the frenzy and rush to the hive entrance, eventually

The primary way colonies reproduce is by swarming during spring and early summer. The old queen leaves the hive with up to three-quarters of the worker force to establish a new colony. The hive has produced new daughter queens, which battle until only one remains alive. In the following spring or early summer, it is her turn to leave the hive and start a new colony. Beekeepers can stop swarming by killing queen larvae. *Photo: Bettina Monique Chavez / Alamy*

with the old queen, and fly to congregate with her on a nearby landmark. After leaving the hive, and from the location of the bee beard, scout bees will hunt for new nest sites.

The first person to begin to understand the way in which honey bees decide on new nest sites was the behavioural scientist Martin Lindauer (1918–2008). When swarming was allowed in the hives by his advisor, the Nobel Laureate Karl von Frisch, Martin would chase the bees through the streets of Munich. He observed how scouts performing these dances on bee beards were often dirty and didn't carry pollen. Instead they might be covered in soot from chimneys or brick dust. They weren't looking for food – they were seeking a new home.[31]

Scouting for a nest site actually starts prior to the swarm emerging from the old hive. The bee scouts have sensed the quality of the site, including its size and entrance size, wind protection, distance from the existing hive, and exposure to the sun. They recruit other scouts to also examine their potential new home. Several potential homes are located and examined, eventually with a large population of scouts showing preference for one of them. The dancing begins again, and the bees frenzy again and excitedly fly off to their new home. Comb construction begins immediately and the queen starts laying eggs within a few days.

One of the leading honey bee scientists of our time, Tom Seeley, describes this process in detail in his book *Honeybee Democracy*, and summarises it well:

This debate [amongst workers for a nest site] works much like a political election, for there are multiple candidates (nest sites), competing advertisements (waggle dancers) for the different candidates, individuals who are committed to one or another candidate (scouts supporting a site), and a pool of neutral voters (scouts not yet committed to supporting a site). Also, the supporters from each site can become apathetic and rejoin a pool of neutral voters. The election's outcome is biased strongly in favour of the best site because this site's supporters will produce the strongest dance advertisements and so will gain converts the most rapidly . . . A unanimous agreement is reached.[32]

What was just one bee colony or superorganism has become two, each complete with workers performing all the tasks and roles needed for a functioning beehive. The mother queen has a new home, which she must quickly prepare in order to survive the following winter. Her daughter, the new queen, is left in the old hive with only about a quarter of the worker force, though this hive has little need of repair or development, and has wax cells ready for her eggs.

Variation in colony reproduction

This pattern of hive reproduction can vary. Typically, only one virgin queen survives the gladiatorial battles in the hive. The hive then splits into two parts: one for the old queen, and a smaller part for the new. Sometimes, however, another queen lives, leading to an 'after-swarm'. An after-swarm forms if there are still large numbers of workers left in the old hive and this old hive contains another virgin queen. Consequently, after-swarms can result in one hive splitting into three. Once a new nesting location has been found and established, the virgin queen will leave her nest site to mate. These after-swarms are typically much smaller than the primary swarm, so they take much more time to build up sufficient numbers to be productive and prepare for the coming winter.

In another variation on queen production within hives, called supersedure, the workers might detect that their mother queen is ageing and her reproductive output is failing. Queen bees can live for four or five years, sometimes even six years. During her tenure the queen has produced pheromones which are transported around the hive after she has been licked and cleaned by the workers. This smell has informed the workers that she is still present and active as an egg-laying machine. When her pheromone and reproductive output finally slows, the hive will produce supersedure queens, often in late summer or early autumn. One to three supersedure cells are produced, though only one of these queens will emerge victorious after killing her sisters. She will fly from the hive, mate, then return to co-exist, at least for a while, with the old queen mother. The queen mother eventually disappears. Most people assume that this disappearance is a result of natural death, although there is some evidence that the workers or new queen may aggressively turn on the queen mother.

Similarly, the sudden absence of those queen pheromones in the hive can cause the workers to raise eggs into virgin queens. Perhaps an incompetent beekeeper has killed the queen in error while working the hive (not me of course, never, almost). In the absence of the queen pheromone, the workers spring into action to save the hive. Only newly laid, fertilised eggs can be selected for this role, before their life's trajectory is inexorably destined by their diet.

Clearly, the workers are in control

The Greek scientist and philosopher Aristotle (384–322 BCE) believed that a beehive was ruled by a king. His ten-volume work *Historia Animalium* ('The History of Animals') is an amazing collection of descriptions and observations of animals and insects, including honey bees. To be fair, his work was incredibly astute for his time. But some of his conclusions were wildly off the mark. In his day, women were considered inferior and nature would never provide weapons to a female, so it was impossible to consider that a hive might be ruled by a queen.

Aristotle couldn't have been further from the truth. Yes, males are produced in the hive, but they don't have any role or opinion that matters and they aren't useful for anything other than a single explosive ejaculation. And yes, in the hive there is typically a single bee that we refer to as a queen. She could be defined as royalty, I suppose, but I hope you'll now agree that it is the female workers who are really in charge. They continually assess the hive and the queen, they democratically decide when to swarm and where to go, and they play a major role in deciding who their future egg-layer-in-chief (queen) will be and when to evict the uselessly alive males.

The beehive is an extraordinarily complex superorganism. A large hive represents kilograms of highly concentrated, protein-rich larvae and delicious honey as carbohydrates. These colonies are a massive resource for us, but also for parasites, predators and disease. The hive is susceptible to and heavily influenced by all of these threats. In the next chapters I will introduce these threats and discuss how they influence the complex behaviour and life history of honey bees. Many of these intricate behaviours are present in other insects, but we know them best from honey bees after studying and befriending them for thousands of years.

2. *VARROA DESTRUCTOR*: THE VAMPIRE MITE

A parasite that gives (viruses) as well as takes (blood and fat)

On 11 April 2000, the parasitic mite *Varroa destructor* was detected for the first time in New Zealand. The hive of a hobbyist beekeeper in Auckland had collapsed and the bees had all died. Thousands and thousands of *Varroa* mites had sucked the 'blood' (haemolymph and fat) of the bees, transmitting disease and eventually causing massive bee mortality and collapsing the hive. The beekeeper immediately reported the infestation to the government hotline for pest incursions. A laboratory confirmed the identification of the mite at 4:15pm on the same day. The then minister of biosecurity later described the incursion as 'probably the most serious breach of our biosecurity in recent times'.[1]

The Ministry of Agriculture and Forestry, the government agency in New Zealand responsible for biosecurity at that time, sprang into action. They carried out a delimiting survey in order to understand the distribution of the mite, moving outwards both north and south from the detected hive. By the end of the first week, the field teams had checked thousands of hives. The boundaries of the infestation of approximately 20 kilometres were established, with a total of 3196 apiaries and 60,479 hives visited over the next few weeks. Just under 10% of the apiary sites examined were found to test positive for *Varroa*. The infested sites were highly clustered around Auckland as a result of the natural spread of the mite, as well as some infestations resulting from the movement of bees used for pollination.[2]

The hobbyist beekeeper in Auckland was probably not the first to experience the devastating effect of these mites in New Zealand. Nobody knows exactly when *Varroa* arrived in the country, nor for how long it had been present. After finding it so widespread, the government speculated that it had likely been present for up to five years prior to its discovery.[3] Similarly, nobody knows exactly *how* it arrived. The importation of live bees into the country has been banned since the 1960s to avoid exactly this sort of problem. Many people I talk to speculate that the mites were hitchhikers on a queen bee and attending workers that a beekeeper smuggled into the country from overseas. But it is also possible that the mites arrived undetected, a little more innocently, in a shipping container.

What next? Should eradication be attempted? Or should we decide that these mites are too well established and widespread for anything but management? After the mite was discovered, most beekeepers wanted an eradication attempt and swift action by the government. That eradication would have meant the destruction of every hive, using petrol for managed hives and laying poisoned sugary baits to massacre feral bee colonies. The National Beekeepers Association supported an eradication attempt, but some members and beekeepers strongly disagreed. There was concern that 'burning the village to save the village' would not work. Thousands of hives would be lost and businesses impacted by a futile attempt to control this pest. People were angry and emotional. One beekeeper indicated he would be willing to 'sit on his hives with a shotgun to prevent government officials destroying them', because they had already concluded that the mite was unstoppable.[4] Sitting on beehives is apparently sometimes an option in New Zealand. *

The debate over what to do next was echoed by the Technical Advisory Group of experts assembled by the Ministry of Agriculture and Forestry. The group was asked three key questions: is eradication technically feasible, what is the probability of success, and what are the risk factors? An initial poll of the 22 group members showed a full spectrum of opinions. Some answered yes, unreservedly, to the question of an eradication attempt. They felt that there was a high probability of success and little risk. Others felt that an eradication was not technically possible. Considerable debate followed, with a focus on the risks.

---

* The phrase 'sitting on a hive' reminds me of another occasion when New Zealand beekeeping made international news. 'RING STING: Hilarious moment idiot beekeeper plants his naked bum onto a BEE HIVE for a £500 bet', reported UK tabloid *The Sun* in 2017: 'Crazy Kiwi keeper Jamie Grainger, 27, lasted a full 30 seconds sat on the angry hive before dashing off in pain . . . He said: "It wasn't pleasant but it was certainly amusing. As you can imagine your a*** swells up. It was just a spur of the moment thing and [a sadistic colleague] offered me a thousand bucks – I was like 'well a thousand bucks', that sounds good."'

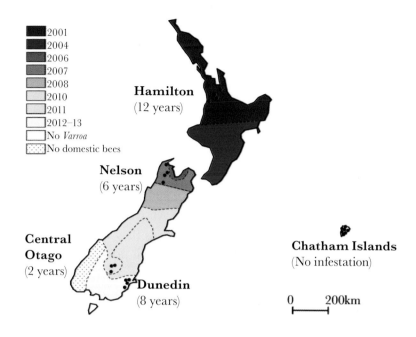

Hamilton
(12 years)

Nelson
(6 years)

**Central
Otago**
(2 years)

Dunedin
(8 years)

Chatham Islands
(No infestation)

0   200km

The spread of *Varroa destructor* through New Zealand, after its discovery by a beekeeper in Auckland in April 2000. *Image: Mondet et al. (2015)*

Non-compliant beekeepers with shotguns were certainly a problem. But infested areas also might not be delimited properly, because the mites are tiny and are difficult to find when in low abundance. And it might be impossible to find and kill all of the feral infested honey bee colonies. Somehow, after considering these problems and more, the Technical Advisory Group concluded that the probability of a successful eradication stood at 17%. After a meeting on 31 May 2000, around six weeks after the first mites were found, the group recommended 'against proceeding with the proposed eradication operational plan because the unresolved associated technical/biological risks mean that the probability of success is unacceptably low'.[5] The government accepted this conclusion, and in July that year announced it had ruled out attempting to eradicate *Varroa*.[6]

Over the next 12 years, *Varroa destructor* moved through New Zealand. It is now everywhere, with the exception of a few small offshore islands.[7] It is a seemingly permanent resident of our country, and of our beehives. In 2020, Australia is the only continent in the world that is free of the devastating effects of this parasite.

Varroa destructor is widely considered the greatest threat to the beekeeping industry worldwide.[8]

The origins and spread of 'the vampire mite'

We have no idea where in the world the New Zealand population of *Varroa destructor* originated. This parasite has a near global distribution and we don't know who brought it in or when it arrived. All we can say is that it didn't come from Australia, which makes a nice change, as they have gifted us many other noxious and invasive species, including brushtail possums, Argentine ants and the plant disease Myrtle rust.

Varroa was initially described in a publication dating back to 1904 as being from Java, Indonesia. The author of the paper, Anthonie Cornelis Oudemans, was a Dutch zoologist who was given the mites to study by the Zoological Museum in Leiden, the Netherlands. Outside the exciting world of mite taxonomy, Oudemans is best known for his publication *The Great Sea Serpent*, which as the name suggests was a review of sea serpent reports from around the world. Oudemans concluded that pretty much all of these ferociously scary creatures were in fact a single, previously unknown, very large seal, which he decisively named *Megophias megophias*, despite the lack of that specimen or any real evidence that it existed. Common names included 'the great unknown of the seas' and 'the Mester stoorworm'. This long-necked seal was 15–30 metres long, with four flippers, a serpentine tail, large eyes and whiskers.[9] Some mocked Oudemans for this description, but mostly because he failed to acknowledge that there could be many more types of sea monster.[10]

Compared with the 30-metre-long Mester stoorworm, the mite that Oudemans described as *Varroa jacobsoni* is tiny. The largest of these mites are 1.1 millimetres long and 1.6 millimetres wide, weighing approximately 0.14 milligrams. The 1904 paper presented *Varroa* as an entirely new genus and *Varroa jacobsoni* as the first species within the classification.[11] Oudemans based his description of this genus only on females, as no males were observed in his studies. But the person who collected the mites must have spent considerable time observing them, as they deduced that the mite was parasitic.

Oudemans went on to explain that the mandibles of the mite 'serve to pierce the less chitinized parts of the bee head and thorax or between thorax and abdomen, to suck juices from the host'.[12] Often found on the floor of the beehive, the mites were described as likely parasites of bee larvae. Oudemans named the species after Edward Jacobson, the naturalist who collected the mites in Java and gave them to the Zoological Museum. *Varroa jacobsoni* is now known as a parasite of Asian honey bees (*Apis cerana*). The mite and its host seem to have co-evolved, and the bee has defensive grooming behaviour that seems to keep mite populations relatively low.

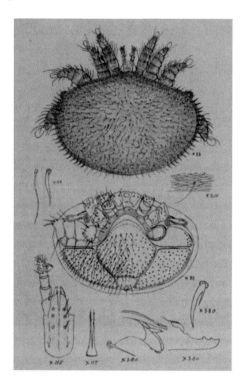

A drawing of a female *Varroa jacobsoni* by A.C. Oudemans.
Image: Naturalis Biodiversity Center / Wikimedia Commons

For nine decades after the discovery and description of *Varroa*, only three species of this mite were recognised. None was *Varroa destructor* and all were from cavity-nesting bees in Asia. Two were collected from the eastern or Asian honey bee (*Apis cerana*), and one from *Apis koschevnikovi*, which lives in the Malaysian rainforest in colonies of around 1000 bees. These eastern or Asian honey bee colonies are often small and flighty, easily disturbed and quick to leave a hive, as well as being poor producers of honey. So, in order to enhance pollination and increase honey production, someone thought it might be a good idea to import the western or European honey bee (*Apis mellifera*) into the region. And, in the early 1960s, *Varroa* was first observed as a parasite of *Apis mellifera* in the Philippines.[13]

Perhaps the mite was first exposed to this newly arrived bee species when Asian honey bees robbed honey from a hive. It's also possible that the mite was spread by beekeepers in the region. Sometimes beekeepers tried to strengthen their *Apis mellifera* colonies by giving them sealed brood from Asian honey bee colonies, which would have been an efficient way of moving any parasites and pathogens between colonies. The Asian honey bee, western honey bee, and *Varroa* mites may have been spread across Russia in this way. By the mid-1960s, beekeepers in this region were reporting counts of over 5000 mites per hive and 70% of the brood infected. Shipments of infected queens and workers were posted around Europe. Japanese beekeepers moved infested bees into Paraguay and Brazil in the 1970s.[14] *Varroa* was first detected in both Wisconsin and Florida in 1987. The world conquest by *Varroa* was well underway.

Given the hive-collapsing effects of *Varroa*, you'd think that people would have quickly realised that they had this pest and done something before mite populations became widely established in a new country. Alas, this is one tricky little pesky parasite. Small populations of *Varroa* are well hidden in beehives. The mite's life cycle takes place within capped cells of the hive, out of the view of your typical beekeeper. On adult bees, the mites are often hidden under the sclerites – the abdominal plates of the exoskeleton. So, if you aren't consciously looking for the mites, you probably won't see them. And after an initial invasion, while the mite population is low, there are often very few effects on the honey bees for several years. Consequently, the mites might be living in a new area for two to six years before hives show any damage.[15]

Varroa destructor: A distinct species

Small things are hard to tell apart and classify. Bacteria are difficult enough. They were first observed in 1676 by Dutch scientist Antonie van Leeuwenhoek, using a single-lens microscope that he designed himself. Leeuwenhoek described bacteria as 'animalcules' and could see only a little diversity. German biologist Ernst Haeckel in 1866 defined bacteria as completely structureless and homogeneous organisms, consisting only of a piece of plasma. Haeckel went on to classify them into the phylum Monera, from the Greek *moneres*, which Haeckel translated as 'simple' but which actually means 'single, solitary'.

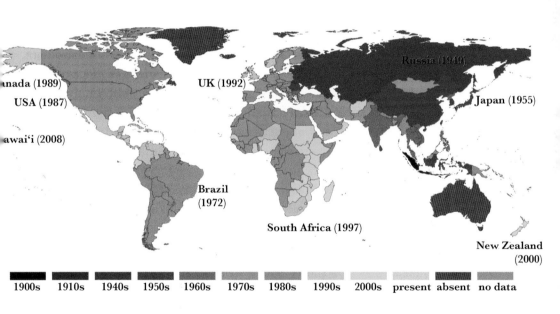

| 1900s | 1910s | 1940s | 1950s | 1960s | 1970s | 1980s | 1990s | 2000s | present | absent | no data |

Varroa global spread, with dates of first discovery. *Varroa jacobsoni* was initially described in 1904 in Java. This map is based on the distribution records as shown in Wilfert et al. (2016). There is some debate regarding the presence of *Varroa destructor* in countries such as Papua New Guinea.

Using genetic techniques, we've now moved on to estimate that there might be a trillion different species of bacteria on Earth – with 99.999% of those species yet to be discovered and described.[16] Bacteria are hugely diverse, and anything but simple.

Genetic approaches using DNA have been of considerable help in classifying life, including *Varroa* species. The first hint that there were more than the three described *Varroa* species came from populations of mites from Java, where Oudemans first described this mite. There, the parasites proved completely unable to infest and reproduce on western honey bees. They were initially considered to be a distinct 'biotype' – somehow slightly different from those mites infesting western honey bee hives. That conclusion led scientists to wonder if there were cryptic variants in these mites, which might mean more than one species is occurring on these bees. Later, a molecular analysis proved this hypothesis correct and populations were found to be genetically distinct. Work published in 2000 first named and described *Varroa destructor* largely based on a genetic analysis. The Australian scientists Dennis Anderson and John Trueman described *Varroa destructor* as 'significantly larger and less spherical in shape than females of *V. jacobsoni*' and also 'reproductively isolated from females of *V. jacobsoni*'.[17] In other words, the two species look a little bit different and, if you put a male from one species with a female from the other, they won't mate. This *Varroa destructor* is the globally widespread and damaging mite you might have in your beehives. Even with a successful control programme, you'll probably have a few mites hiding somewhere.

It was just 60 years ago that *Varroa destructor* was first exposed to the western honey bee. It has had a very short history of evolution with its delicious bee hosts, which are clearly to the little vampires' liking. On the other hand, Asian honey bees have a long evolutionary history with these mites, leading the bees to develop a suite of hygienic behaviours against *Varroa* parasitism.

Hygienic behaviours in social insects are diverse. I've talked about some of these already, such as the way honey bees encapsulate mice with propolis to inhibit disease. Honey bees also display 'social fever', whereby workers attempt to increase the temperature in the brood patch in order to kill pathogens, effectively cooking them. They might also aggressively shun and avoid infected adult bees to reduce the transmission of disease. These are all examples of social immunity. Social immunity in bees typically involves behavioural, physiological and organisational traits that enhance the health of the colony. Individual bees might suffer by being shunned, but overall the colony benefits.

Asian honey bees have evolved very specific social immunity and hygienic behaviours towards *Varroa*. They are able to quickly detect infestations and remove infected brood at the pupal stage, thereby reducing or preventing the population growth of mites within a hive. This behaviour is given an array of terms: 'removal response', 'removal behaviour', '*Varroa*-specific hygienic behaviour', 'suppressed mite reproduction', and, my favourite, '*Varroa*-sensitive hygiene'. Other hygienic behaviours of Asian honey bees include the 'entombing' of infested drone brood cells beneath propolis, resulting in the non-emergence and eventual death of both infested drones and parasites. It's a slow death for the poor male, but a small sacrifice for the greater good of the colony. And it's no big loss, really, as males are useless around the hive. Asian honey bees also engage in aggressive grooming, which involves physically removing and destroying adult mites from adult bees. The mites are chewed by the infested worker herself or by her nestmates.

It makes sense that Asian honey bees have developed these hygienic behaviours to *Varroa* parasites, given their long co-evolutionary history. And it makes sense

Two *Varroa destructor* mites on a honey bee pupae. The bee pupae was sealed in a cell on a bee comb in a hive. *Photo: Phil Lester*

that western honey bees struggle with *Varroa*, with hives often succumbing, since the mites and bees have only recently been brought together.

It's now the case that if a beekeeper with western honey bees has *Varroa destructor* and does nothing to control it, their hives will experience massive rates of mortality. Some scientists go even further, such as pollination biologist David Pattermore, who argues, '*Varroa* means that, for all practical purposes, honey bees only survive now thanks to human intervention in managed colonies.'[18]

Varroa destructor has been described as *the* major pest of honey bees. It is considered the crucial factor in the decreasing numbers of honey bee colonies, and beekeepers, in Europe. It is predicted that, together with the worldwide decline of natural pollinators, *Varroa* will only compound future pollination problems.[19]

The weird life of *Varroa destructor*

A massive amount of work has been done on *Varroa*, given its economic and environmental importance. To learn about the life cycle of the mite, scientists have constructed small windows in hives so that they can watch mites develop on parasitised bees. I'll attempt to summarise what we know about *Varroa* here.

This little blood-sucking parasite is inextricably linked with its host. There is no free-living stage; it is never more than a few millimetres away from a honey bee. Most researchers believe that if a mite is ever away from a bee it will live for only a few hours (although I've heard of one scientist who is rumoured to have kept *Varroa* mites on his desk, where they lived for about three weeks cannibalising each other). There are two phases of a mite's life cycle: one phase within a single cell of a juvenile bee, and another where the mite moves between cells or hives on the adult bee.

First, an adult female *Varroa* mite rides around on an adult bee. The bee visits many different cells within the hive. She checks on young larvae, with no apparent response from the mite. But when the bee tends or passes by older larvae that are nearing pupation, her passenger springs into action. At the late developmental stage, called the fifth-instar or growth stage, larval honey bees give off a distinct smell. Adult female *Varroa* are able to recognise that a bee is at

1. A queen bee lays an egg in a brood cell. During peak production she might lay 1500 eggs a day.

2. Worker bees feed the larva. When the larva reaches the fifth instar, it becomes attractive to *Varroa*. A mite leaves the worker bee, enters a cell and hides in the larval food.

3. Once the cell is capped, the mite cuts a feeding hole in the pupa. She lays her first egg, which becomes a male. She lays up to five more eggs, all of which are female and mate with the male.

4. The mite lays an egg every 30 hours. She can lay eggs in several bee cells in her lifetime. Mite populations can double every four weeks.

5. The mite feeds on the fat body of the bee and transmits viruses. The emerging bee may have crippled wings, reduced foraging ability and a shorter life span, and may be more vulnerable to disease and pesticides.

Illustration: Bayer Scientific

this stage. It must be an odour that is attractive to the mites, because they have no eyes and are blind – and it is dark in the hive. Scientists have spent huge amounts of money trying to identify this 'host odour', without success. If we knew what attracted the mites to fifth-instar cells we could synthesise an artificial chemical and trap the mites – but to date no one has identified that perfume.

It is this fifth-instar bee that the mite has been patiently and cleverly waiting to find. But then the *Varroa* mite seems to do something particularly stupid. She crawls off the bee, passes between a larva and its cell wall, and immediately becomes stuck and submerged in the larval food at the base of the cell.

Perhaps, however, there is logic to the mite's behaviour. Submerged in the food, she is effectively hidden from workers who might overwise register her presence. She has a little snorkel-like organ that allows her to breathe through the food. Our vampire mite will stay there until the cell is capped by worker bees and the larva has consumed all the food. With the larval cell capped, food eaten and the larva undergoing pupation, the mite is now free to attack its host. The adult female mite cuts a hole in an abdominal segment of the pupa cuticle and feeds on the bee. This single feeding zone is needed because the little mite offspring that will arrive in the next few days cannot penetrate the bee exoskeleton by themselves. It's a nice example of parental care in the mite world.

Varroa could be described as tidy but gross. The adult female deposits her faeces in one particular spot on the honey bee cell – above the bee and towards the cell base. This is known as the faecal accumulation site. The mite becomes relatively lazy at this time. She sits on her faeces, only occasionally descending to feed on the bee or venture along the cell wall to lay an egg.

The female mite's physiology has been changing ever since the honey bee cell was capped, with her eggs and reproductive system beginning to mature. About three days after the cell is capped, her first egg is produced. *Varroa* have a haplodiploid mating system, with unfertilised eggs becoming male, just as with honey bees. Her first egg is unfertilised and will become a male. Subsequent eggs, produced at approximately 30-hour intervals, will be fertilised by the adult female. These eggs are diploid (with two sets of chromosomes) and will become females. A normal reproductive programme for the mite will result in up to five eggs being produced in worker bee brood cells, and up to six eggs in drone brood cells. The male mites are much smaller than the females and take just under six days to grow into adults. The female mites take just under seven days.

Once a male reaches maturity, he finds his way to the faecal accumulation site that his mother started. And there he sits and waits for his sisters to mature.

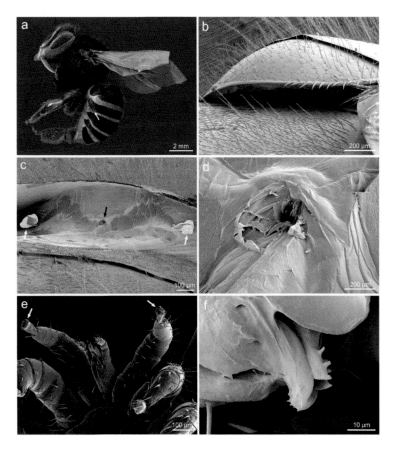

This microscopy figure shows *Varroa* on an adult honey bee. (a) Mite's location shown with an arrow. (b) Mite is hiding under the bee's exoskeleton. (c) Mite has been removed. White arrows show where its foot pads were clinging to bee; black arrow shows puncture location. (d) Puncture site wound. Mite's mouthparts (green and yellow) are still embedded in bee's skeleton. (e) Front feet and mouthparts of mite. (f) Another view of wound site, with mouthparts. *Images: S.D. Ramsey et al. (2019); see endnote 24.*

He will smell his sisters coming – young virgin females produce volatile sex-pheromones that rouse their brothers. Excitedly, he cleans his chelicerae, the pair of appendages at the front of his mouth. The first female matures and she too makes her way to the faecal accumulation site. This location now becomes the 'I'll have sex with my sister' site (though you won't find that terminology in the published literature). Once his sister arrives, the male mite removes a package of sperm from his genital opening using his chelicerae. He carefully places that sperm package in her genital opening. That process is repeated several times in order to fill her spermatheca, the sperm storage organ. The sperm are immature but will complete their development within the female, where they are nourished and will live for as long as the female mite herself.

As each female ages, she produces fewer and fewer sex-pheromones and the male finds her less and less attractive. Consequently, he focuses his attention on the youngest and most fragrant female in the cell, until the next attractive young individual completes her development into a voluptuous adult. The number of females that complete development and mate with the male depends on the time it takes for the honey bee to pupate. Male bees take longer to develop, so more mites will complete their development on drones, whereas worker bees take less time to pupate, so fewer adult mites are produced on those. The average number of new female mites to emerge from worker cells is 1.45, with the remainder dying before they can mature.[20] So the mother mite can successfully raise only one or two daughters on worker bee cells. But if the cell contains a drone bee, *Varroa* in this cell will produce an average of 2.2 new mites.[21] That sounds like a low rate of reproduction, but each adult female mite can go on to invade and reproduce in up to seven different honey bee cells.

Mites that have not completed development by the time the bee exits the cell will die. After the bee and fertilised adult female mites leave the cell, the male dies too. His work is done.

Queen larvae are not attractive to these parasitic mites, because queens emerge at 16 days. That's a whole five days faster than a worker and would leave no time for any daughter mites to successfully mature into adults. Hence, the *Varroa* mite avoids queen cells, probably due to a specific chemical odour emitted by the queens or their food.

The next stage of the mite's life cycle begins. The mother mite and any daughters that have managed to complete their development into inseminated adults begin the hunt for new bee larvae. A middle-aged nurse worker bee is the ideal transport mechanism for *Varroa*. The parasites are a bit like Goldilocks, with her fussiness for porridge: a young nurse is less appealing, an old one not quite so good, but most of the middle-aged nurses are just right, and so they will have the most mites in the hive. The mites seem able to recognise the age of a bee from its hydrocarbon odour. They ride around the hive on the nurses, looking for new hosts, abandoning this ship and surreptitiously slipping into a fifth-instar bee cell when the time is right.

Beekeepers and some of the first scientists working on *Varroa* observed that drone brood had a rate of infection up to ten-fold higher than that of workers. The mites seem to prefer drones to worker bee larvae. Given that *Varroa* seems able to distinguish middle-aged nurses from younger or older bees, perhaps the mite can actually smell out and preferentially select a male. This preference would be to the mite's advantage, because drones take so much longer to pupate and develop. Consequently, the female mite is able to complete the development of many more of her daughters on drones than on female workers.

Varroa on a bee. These mites reach their highest abundance on drone brood, because drones spend longer in the pupal stage than workers do. *Photo: Phil Lester*

There could, however, be other reasons why drones collect more *Varroa*. For example, worker larvae are attractive to *Varroa* and are infested 15–20 hours prior to cell capping, whereas drone brood are attractive for between 40–50 hours. Drones are thus available for infestation for twice as long as workers. Drone brood are also substantially larger and require more feeding and attendance by nurses, providing more of an opportunity for the mites to exit from a nurse and slip into a drone cell.

Movement between hives is the next challenge for the mites. The adult female mites are already riding around on workers and drones. Those individuals will leave the hive and periodically get lost, perhaps because they are ill from the *Varroa* infection, and may drift into other hives. Eventually the *Varroa* infestation weakens an entire hive. There will be fewer workers and fewer guards at the hive entrance. A weak hive that was once strong represents a tempting bounty for any neighbouring bee colonies. The large stocks of honey and few guards make for easy robbing. While the neighbours are helping themselves to the pantry, mites climb aboard and are whisked away to a new hive and population of bee larvae. Some beekeepers refer to this scenario as '*Varroa* bombs'. Perhaps you are a diligent beekeeper and work hard to control your mites, but perhaps your neighbouring beekeeper isn't. If your strong colony robs from your neighbour's weakened hives, the mites in your neighbour's colony become your problem too. Suddenly your hive accumulates a substantial amount of honey – and a massive mite problem.

What happens in winter, when no larvae are being produced? The adult female mites ride around on and feed off the winter bees. *Varroa* have been shown to live on adult honey bees for 100 days in the absence of brood. They lie in wait for the spring bee bloom.

The fat-sucking effects of *Varroa*

Why is *Varroa* so devastating for beekeepers? Let's work through the effects of this parasite, and the diseases it carries, for an individual bee. Then I'll discuss its effects at the colony level – what happens to your entire hive.

A parasite can cause a lot of damage as it consumes a living thing, especially if that organism is still growing and developing. We see this in human populations. For example, a 1999 report noted that 9 out of 10 children in the Philippines are prone to poor physical and mental development, likely due to intestinal worms. In children, infection by these worms causes stunted growth, decreased physical activity and poor physical and mental development.[22] The rate of infection in children and the long-term effects must affect the functioning of the entire country.

We see remarkably similar effects in honey bees after *Varroa* infection. Bee pupae that have mites feed on them will have lower body weight as adults. Drones might be as much as 19% lighter than their non-infested brothers. Adult bees parasitised as juveniles also have a reduced lifespan. They start foraging earlier. They have a reduced ability to learn, which is probably associated with a relatively poor ability to navigate. This results in prolonged absences from the hive and a low rate of return altogether.[23]

In the title of this chapter I described the mites as feeding on 'blood and fat'. Recent research has indicated that is an accurate description of the *Varroa* diet, but it would be more accurate to say they seem to eat almost entirely fat with a little blood on the side. Specifically, the mite seems to attack and consume a honey bee organ called the 'fat body'.[24]

We can see that mites are fat-body feeders by looking at their digestive systems. This parasite's gut is very different from that of a blood-sucking invertebrate. It is more like that of a species that externally digests their victims. These animals secrete enzymes into their host's body and suck up the partially digested contents. Most spiders eat and digest their prey in this way. The closest mite relatives to *Varroa* also exhibit this sort of behaviour. When feeding, the mites don't deposit or remove the excess water that is common in blood-feeding parasites; instead, the composition of their faeces is consistent with direct organ feeding. Finally, we know that the mites feed primarily on bee fat bodies because of studies using microscopy, in which the fat and haemolymph of bees are effectively dyed with chemical colouring. An adult mite's preference for nurse bees now makes further sense, as a 2019 study notes: 'Nurse bees have substantially larger and, ostensibly, more nutritionally dense fat body than other stages of the worker bee caste'.[25] If you are an adult mite wanting the juiciest, tastiest meal, choose a middle-aged nurse.

So, okay – the mites primarily feed on fat bodies in honey bees. So what? What's a little bit of liposuction? I've considered such a procedure myself as the middle-aged spread and dad jokes become commonplace. But, where I could stand to drop a few pounds for the benefit of my health, fat-body feeding is a big health problem for bees. The fat body of an insect performs a wide variety of tasks. It is the primary storage and synthesis site for protein and lipids. If the fat body is damaged when the bee is a larva, the bee is less able to store protein from the pollen it consumes as an adult. This organ has a major function in the bee's immune system, especially in fighting bacterial and viral infections. The fat body helps regulate metamorphosis, metabolism and thermoregulation, and is critical to overwintering success. The fat body also plays a big role in pesticide detoxification. There are many similarities of function between the fat body and the human liver.

In short, if there were one organ in the bee body that you *wouldn't* want a parasite to feed on, it is the fat body. Feeding on bee blood would probably be less damaging. The diverse roles of the fat body are likely why the effects of

An adult honey bee with stunted, useless wings, after infection as a juvenile with the deformed wing virus. These bees are often ejected from the hive and left to die. *Photo: Phil Lester*

Varroa interact, and worsen, in the presence of other stressors like pesticides, pathogens and food shortages. A bee's ability to cope with these external stressors is substantially reduced.

Perhaps the most obvious effects of *Varroa* parasitism is a honey bee with deformed wings. Her crippled wings mean that she will never fly. She has a shortened abdomen and is frequently less hairy than healthy bees. She looks sick. Her symptoms are associated with the mite feeding on her, as well as the deformed wing virus (DWV). I'll cover viruses in detail in the next chapter, but it is clear that the parasitic mite and DWV are linked in a mutualistic symbiosis – they live together and help each other. The mite acts as vector of the virus. The fat-body feeding of the mite helps suppress the bee's immune system. The virus can further suppress the immune system, which in turn supports mite feeding and reproduction, and results in the virus spreading even more.[26]

DWV isn't the only virus or disease to be spread by *Varroa*. We've known for over a decade that at least a dozen different viruses are associated with this parasite.[27] Kirsten Traynor from the University of Maryland describes the interaction well: 'The mites are basically dirty hypodermic needles. The main diet for the mites is blood from the developing bee larva. When the bee emerges, the mites move on to the nearest larval cell, bringing viruses with them.'[28]

In 2016, at the time Kirsten was quoted, we didn't know about the external digestion displayed by *Varroa* on their bee hosts. We now know that those dirty hypodermic needles are largely used to consume bee fat bodies, but the results for pathogen transfer are much the same, so her description is still very apt.

Varroosis and the death of your beehive

What are the effects of a *Varroa* infestation on your entire beehive? You might think that the effect on the colony is just the sum of the effects on the individuals. Unfortunately, it's far from that simple. Part of the complexity is due to your honey bees forming a 'supercolony'. If some individuals are failing at one task, such as foraging, others will step up and change their role to satisfy this need. The bees also display the social immunity suite of behaviours and biophysical defences to infection by *Varroa* and other parasites and diseases. Sick larvae and adults are physically ejected or removed from the hive.

It is clear, however, that typically the supercolony beehive has a limited ability to buffer the effects of *Varroa*. Too many mites and too much parasitism at the wrong time will mean the supercolony dies. The culmination of *Varroa* and the diseases it carries is referred to as varroosis. The impact of varroosis on the hive is related to brood production and the number of adult bees; for example, an abundant production of drone brood can enhance the population growth of mites.

In the New Zealand summer, what can appear to be a strong hive in late January might be near empty, with only the queen and a few dozen bees, at the end of February.[29] The hive will never recover. A similar, common pattern

When *Varroa* reach high densities in the hive, the brood can show Parasitic Mite Syndrome. Symptoms can be observed before and after the cell is capped. The brood pattern will be spotty, as in this photo. Bees may not be strong enough to crawl out of their cells, and will die. Mites can be seen on these bees, and some bees have deformed wings. The symptoms can be confused with American foulbrood. *Photo: Frank Lindsay*

is observed from *Varroa* infestations in Europe.[30] In mid-summer the bees and brood appear relatively healthy, despite the presence of some mites. The ratio of brood to adult bees then becomes higher than in *Varroa*-free colonies, which can be a sign that the colony is at risk of future collapse. Near the end of summer, the clinical symptoms of DWV appear. The ability of the honey bees to learn and to forage decreases. By autumn, the physiology of the bees has been strongly affected, with lower bee weights. You'll remember from chapter 2 that these winter bees are protein and fat stores for the colony. They are a special caste of bees that are essential to the colony's chances of surviving the winter. If they are in poor health, the colony is in trouble. The hives will collapse in autumn and will continue to die through winter and early spring.

Your hive might also experience Parasitic Mite Syndrome (PMS) under heavy infestations of *Varroa*. PMS is associated with *Varroa* but has a range of variable symptoms. The brood are patchily distributed, with shrunken and perforated caps. The larvae assume an odd, twisted position, slumped in the bottoms or sides of their cells, and the pupae have an odd yellowish colour. Bees might try to remove larvae or pupae from cells, sometimes successfully, giving the impression that the diseased and dead immature bees have been 'chewed down'. We don't yet know what causes PMS, other than that it is associated with *Varroa* infestations – and mite control might even cause these symptoms to disappear. It is certainly possible that PMS is caused by a range of pathogens interacting with *Varroa*.

Traditional and emerging approaches to managing *Varroa*

Ideally, with any method of managing *Varroa*, you should first monitor the mite populations. Every expert on *Varroa* management will tell you to monitor, monitor, monitor.[31]

My favourite technique for monitoring *Varroa* abundance is called an 'ether roll'. I've never actually used it, but this approach sounds like it would be fantastic fun. The first step is to somehow get yourself an aerosol can of ether. Ether is an extraordinarily flammable chemical even when it isn't sprayed; it is even more flammable in aerosol form. I can't imagine how often fires and flames must erupt as beekeepers spray the ether near the burning material in smokers that are used to calm bees. To use this aerosol ether, you put some bees into a jar and spray the ether onto them. Mites fall off and are stuck to the sides of the glass. Then you stick your head in the jar and count the mites. Bear in mind, ether is an anaesthetic and was frequently used to knock people out for surgery. I wonder how many beekeepers are well-rested after a day counting *Varroa* using ether.

There are a range of other, probably less enjoyable approaches to monitoring *Varroa*. You can use an alcohol or soapy water wash, or a 'sugar shake'. People also do visual inspections of bees (which is hard if the mites are hiding in the bee exoskeleton) or count the number of fallen mites on the base-boards of the hive. For a sugar shake, you collect bees into a jar with icing sugar. After rolling the bees in sugar, you put the mixture through a sieve. The mites are dislodged from the bees and can be counted. Most bees will survive, and when cleaning themselves will have a nice sugar rush. We don't know why the sugar dislodges the mites, though perhaps it interferes with their ability to hold on to the bee. Or perhaps the sugar makes the mite want to groom itself, so it lets go of the bee.

It's especially important to monitor mite populations when using chemicals to control *Varroa*, as part of an effort to limit the mites' resistance to pesticides. There is a long list of chemicals used to control this parasite. Some are considered 'organic' and others 'synthetic'. The first mainstays for mite control were the synthetic chemical fluvalinate, sold as Apistan, and flumethrin, sold as Bayvarol. Both are pyrethroid-based pesticides. Extracted and developed from flowering plants in the *Pyrethrum* genus, pyrethroids are a widely used class of insecticide that can kill honey bees. Apistan and Bayvarol are delivered and used in concentrations that primarily affect mites. Unfortunately, *Varroa* have shown resistance to Apistan, which seems likely to worsen with continued use. In the North Island of New Zealand, *Varroa* is also rumoured to be showing major signs of resistance to Bayvarol. One beekeeper I talked with thinks that 'both strips [Bayvarol and Apistan] should be withdrawn so we can use them every second or third year, for one spring period'. This would likely extend the efficacy and use of these pesticides over the long term.

'Organic' treatments include essential oils and organic acids. Essential oils are plant-derived and highly volatile, including oils from plants like thyme and eucalyptus. The organic acids are primarily formic acid, lactic acid and oxalic acid. Though they can be considered organic, they are nevertheless potentially hazardous to bees and beekeepers. The key issue is that these acids are vaporised, and vaporised acid is frequently about as nasty as it sounds. As a general rule, you should try to avoid inhaling acid. Despite such inhalation issues, many beekeepers I've spoken to efficiently control *Varroa* in their hives with a combination of careful monitoring and the use of these organic acids.

Varroa is a massive, global problem for beekeepers. A commercial beekeeper might pay NZ$50,000 per year in treatments,[32] with developing resistance as a major issue. Consequently, the search is on for new organic and synthetic ways to control these parasites. To date, all methods have issues. Pesticide resistance, health and safety issues, residue effects or tainting the honey are common problems. There is a lot of money to be made if a company finds a new and efficient way of controlling these pests.

Novel and emerging approaches in this field include the use of 'gene silencing' techniques. In 2012, scientists identified that gene expression in *Varroa* can be modified by feeding bees double-stranded RNA (dsRNA).[33] The bees ingest the dsRNA and pass it on to the mite when it feeds on the bee. This dRNA can be designed specifically with the mites in mind, inhibiting gene function within the mites and causing substantial mite mortality. That's a fantastic idea and result. As it is specific to the mite, this dsRNA doesn't harm bees, beekeepers or other plants or animals. It is sometimes called 'gene knockdown'. I think these sorts of techniques have immense potential and should be a priority for the next generation of 'pesticides'.

A step further in this approach is the use of genetically modified bacteria. In January 2020, a group from the University of Texas published research on a novel and apparently effective method of *Varroa* control. Their work is preliminary but offers hope. Sean Leonard, a PhD student at the University of Texas, is developing tools and methods to genetically alter the bacterial community that lives in the honey bee gut. In his 2020 paper in the journal *Science*, Leonard described a process whereby *Snodgrassella alvi*, an already beneficial species of bee gut bacteria, was genetically modified to produce dsRNA.[34] This dsRNA then interferes with or silences gene expression in a highly targeted manner. Leonard showed that bacteria could be modified to produce dsRNA that would help bees resist a viral infection. He and his team produced another strain of the bacteria that produces the dsRNA that specifically affects *Varroa*. When the mites feed on the bees, they also consume the dsRNA. Leonard genetically modified the bacteria to stop one of the mite's vital genes from working, resulting in substantial mite mortality.[35] These tools show that microbiome engineering has real potential to increase pathogen resistance.

This work has been described as a potential 'microbiome silver bullet for honey bees'.[36] The authors and others note, however, that there is still much work to be done. Not least of which involves the ethics of genetically modifying organisms for release in the big wide world. Countries such as New Zealand strictly regulate the importation, development, testing and release of genetically modified organisms. These questions and many others need careful consideration.

Honey bee researcher Jana Dobelmann monitoring for *Varroa* in her hives using the 'sugar shake' method. *Photo: Phil Lester*

Breeding bees that display *Varroa*-sensitive hygienic behaviour

As we've already seen, Asian honey bees cope with *Varroa* infestations using a range of hygienic behaviours. They quickly find *Varroa* and remove infected brood. They 'entomb' infested drone brood cells beneath propolis. The drone is then trapped inside his pupal cell, from which he and his parasites will never escape alive. The numbers of mites in Asian honey bee hives are typically low. Is it possible to find and select for similar hygienic behaviours in the western honey bee, despite it not co-evolving with these parasites? Many people think the answer is yes – hygienic behaviour can and already does occur in the western honey bee.

Evidence for some bee colonies offering resistance to *Varroa* has been observed in many countries. Due to new molecular genetic approaches, there is now evidence for at least 73 different genes related to hygienic behaviours in bees.[37] A honey bee's antennae (the bee equivalent of a nose) seem to play a major role in *Varroa*-sensitive hygienic behaviour. Bees that are hygienic have a greater sensitivity via their antennae to the deformed wing virus or *Varroa destructor* virus infections.[38] That's all very logical: for a worker bee to do something about this parasite, it first has to detect it. These studies offer a pathway forward for selecting genotypes and following different genes for hygienic behaviour in a bee population. It's very much the 'academic' or 'scientist' way forward. Of course, us researchers would also want to take that approach.

But another way forward uses a blunter tool that could possibly lead to faster results. Beekeepers all over the world have correctly guessed that artificial control of *Varroa* by the use of chemicals inhibits any natural selection for resistant bees. What if they stopped using pesticides? Bees that do not display *Varroa*-sensitive hygienic behaviour would all die or could be weeded out, leaving only those with these highly desirable traits.

This tactic is based on the idea of letting *Varroa* kill off hives without *Varroa*-sensitive hygienic behaviour and then the beekeeper breeds from those that survive. Gary Jeffery, a beekeeper on the West Coast of the South Island, took this exact approach. His hives were devastated in 2014, when *Varroa* killed 99.4% of his colonies. Of 1000 hives, he was left with just six. That level of mortality would be enough to drive most people out of the business. However, Gary has spent the last several years breeding from those six hives, without using chemicals or mite control. 'From the healthy six hives showing resistance we filled 300 hives over the past four years and at the same time researched to breed more bees resistant to the *Varroa* mites. We do have hives now that do not require miticide treating to keep them healthy,' he has said. Gary describes his hives as getting 'one or two' mites. But they do not seem to suffer from varroosis or Parasitic Mite Syndrome. 'We are obviously biased but we think we now have the perfect honeybee. They are very good workers, very quiet to handle and not affected by the common bee diseases.'[39] Gary's bees might well prove to be *Varroa* resistant and display some sort of *Varroa*-sensitive hygienic behaviour. And, if so, these bees would be a huge benefit to beekeeping and agriculture both nationally and internationally. His work has been hampered by a lack of funding, but I'm glad to say that others are trying to help, including myself.

Several strains of bees have been developed in this fashion, with high hopes for *Varroa* resistance. 'Russian (Primorsky)' bees, for example, were renowned for their resistance, but ultimately they were not recommended because they also displayed a reduced brood and honey production.[40] More recently, there have been positive signs that honey bees are evolving social immunity or a form of behavioural resistance to *Varroa*. Researchers in Europe have found honey bee strains that detect *Varroa* within capped cells.[41] After detection, the worker bee simply uncaps the cell and then, without removing the developing bee or the parasite, reseals the cell. This is a really, really clever behaviour. *Varroa* are very sensitive to any slight changes in temperature, pheromones or humidity. The mite's reproduction can be irreparably damaged simply by an adult bee opening the cell and subtly changing the environment. Opening the cell damages the mite and has the additional benefit of not killing the valuable developing young bee.

The authors of this study conclude that 'aiming at sustainable global apiculture, it appears prudent to employ evolutionary thinking to manage infectious diseases by taking advantage of efficient mechanisms favoured by natural selection'.[42] What they are suggesting is that a focus on *Varroa* control using pesticides is not the appropriate approach. Pesticide use might even inhibit mite control over the longer term, because we'd be selecting for pesticide resistance in the mite and inhibiting selection of social immunity in bees. Instead, letting *Varroa* kill off susceptible hives and breeding from those that are resistant is the better approach. We humans can control the evolution and natural selection of this bee and parasite relationship. Our management approach will direct their populations.

One final note of caution. By taking the approach of using no pesticides, letting *Varroa* go unchecked, and breeding from the few surviving hives, you might not always be selecting hygienic behaviour or social immunity. You might be selecting something else altogether. A classic example is the story of Ron Hoskins' bees.

Ron Hoskins, an 89-year-old beekeeper from Swindon, England has lost hundreds of hives due to *Varroa*, which entered Britain in 1992. He spent 18 years selecting and breeding bees that were resistant to the effects of these parasites. Without pesticides, susceptible colonies would die, leaving only those that displayed some resistance. He had success, and bred what have been described as 'super bees' that don't need mite control. It turns out, however, that Ron's bees were hyper-infected with a strain of the deformed wing virus that is relatively benign. The nasty viral strain that is associated with *Varroa* and varroosis cannot get a foothold on these hyper-infected bees. Virologists call this process 'viral superinfection exclusion'.[43] Ron was selecting for the virus, rather than for the bees. Unfortunately, it is thought that simply moving Ron's bees to another apiary would be futile. Researchers think that in different locations his bees would likely be swamped by the nasty viral type carried by *Varroa*. The bees would face the disease common to most colonies, and would likely suffer high mortality.[44] * I'll discuss these viral strains and this study in the next chapter.

* I've a sad update on Ron's bees. In March 2020, his apiary workshop and many hives were destroyed by vandals. He had spent 20 years selectively breeding these bees and researching their *Varroa* resistance. His work was known around the world. Many beehives were saved, but Ron was reported to be devastated. People can be awful. A donation page was set up, raising more than £17,000 within 24 hours. 'It's wonderful but I still hope whoever did this got stung and someone reports them,' he told the BBC.

Below is a section from the review 'Biology and control of *Varroa destructor*' by Peter Rosenkranz and his team. It was published in 2010 but remains very relevant today:

> *Varroa* mites have been considered a problem for beekeeping for about 40 years . . . So, we now look back on more than 30 years of intensive research on various aspects of the biology pathology and management of this parasite. To summarize the efforts we can state that we have significantly increased our knowledge on mite distribution, pathogenesis, host–parasite interactions and effective use of certain treatments. In most countries the *Varroa* situation is stable; the beekeepers have learned to 'live with the mite' and most of them do not know beekeeping without Varroosis. Most extension services of state experts and bee-keeping organizations have successfully focused on integrated *Varroa* management considering the local beekeeping conditions.
>
> However, we must also state that we have not achieved the original aim to get rid of the parasite or at least to solve the problems related to Varroosis. There is neither a *Varroa* treatment available which fulfils all the criteria 'safe, effective and easy to apply' nor a honey bee which is sustainably tolerant to Varroosis under temperate climatic conditions. Rather, we now face new problems with secondary diseases and damage in honey bee colonies caused by synergistic effects of Varroosis plus other pathogens or environmental factors. In addition, there are still no data showing that *Varroa* in general becomes less virulent or that honey bee colonies selected for mite tolerance survive without mite control. These aspects will maintain the pressure on honey bee colonies and beekeepers especially in the non-tropical countries with the significant risks for pollination services.
>
> This also means that further *Varroa* research is urgently needed . . .[45]

Since that review was published we have made further advances, but mostly in our understanding of the biology of *Varroa* and not in its control. I am hopeful that 'super bees' or even 'the perfect bee' might be in the making. Governments, scientists and beekeeping industry leaders urgently need to work together to control this globally leading cause of honey bee mortality. We also need to hear from the Gary Jefferys and Ron Hoskins of this world – those people who patiently and carefully work in the background, often for decades, with the potential to make major contributions to bee health and the apicultural industry.

We should keep an open mind, too, to approaches such as gene silencing with dsRNA. These are potentially effective control methods for this hugely important parasite.

Gary Jeffery tending a beehive with his granddaughter, Renee. A beekeeper for 70 years, Gary believes he has bred the 'perfect honey bee'. *Photo: Phil Lester*

3. VIRUSES

Manmade, global pandemics in our favourite insect species

You've probably had the flu (influenza, from the influenza virus) before, or sometime in the next few months you'll catch a common cold (frequently caused by a group of viruses called rhinoviruses). I hate having the flu. For about a week I have a fever, a headache and muscle pain, and my bones seem to ache. My wife, Sarah, lovingly fusses over me doing everything in her power to help. She drives me absolutely nuts. I'd much rather miserably suffer through my sickness in solitude, wallowing in my self-pity alone.

The annually different strains of the flu have different effects, some more severe than others. There have been three major global flu pandemics in the twentieth century; the current coronavirus pandemic occurring as I write this chapter is quickly becoming another. The most infamous of these historic pandemics is the outbreak of Spanish flu (also known as the 1918 flu pandemic) after the First World War. It affected at least 500 million people and killed 50 to 100 million people (3–5% of the world's population at the time). It was described as the 'greatest medical holocaust in modern history'.[1] To New Zealand's great shame, we introduced this virulent flu strain to the Pacific nation of Western Sāmoa. On 4 November 1918, a ship travelling from Auckland docked in Apia carrying six seriously infected passengers. Twenty percent of the Samoan population died after catching this strain of the flu from one or more of the passengers.[2]

Fortunately, the regular strains of the flu that annually sweep the world's populations – typically starting in Asia and following winter around the globe – cause much less havoc, though they do still kill. Common colds are much less nasty. Over 200 different virus species can result in a running nose and a cough, with adults typically getting two or three colds a year. These colds are usually not serious enough to stop you from going to work and generously shedding and spreading these viral infections amongst your colleagues.

There are two points worth making from this detour into human flu and cold viruses. Firstly, there is a massive diversity of viruses, and they can cause a wide variety of symptoms in their hosts. Many viruses can be relatively benign. You might not even know you host them, with a very small proportion causing a few sniffles that will slow you down just a bit. Their effects might become more prominent if there are other stressors such as over-tiredness, an immune system compromised by other illnesses, or a poor diet. Other viruses might just kill you no matter how much sleep you get. Secondly, within a viral 'species' we occasionally see variants or strains. Some strains do little, while other strains cause global pandemics with large rates of mortality. These pandemics are often aided by human movement and migration, whether the virus parasitises humans or honey bees.

As discussed in the previous chapter, an example of a perfect, worldwide viral pandemic is the deformed wing virus (DWV) associated with the *Varroa* mite. The global spread of this virus was recently described in a review led by Lena Wilfert of Exeter University as a '*Varroa*-vectored virus pandemic'. The authors describe how the transmission of DWV between bees is normally inefficient, but *Varroa* facilitates and enhances transmission:

Deformed wing virus exhibits epidemic growth and transmission that is predominantly mediated by European and North American honeybee populations and driven by trade and movement of honeybee colonies. Deformed wing virus is now an important re-emerging pathogen of honeybees, which are undergoing a worldwide manmade epidemic fuelled by the direct transmission route that the *Varroa* mite provides.[3]

This virus, and particularly the strain transmitted by *Varroa*, is nasty.

What actually *is* a virus?

The English word 'virus' comes directly from the Latin *virus*, meaning 'poison'. Viruses are relatively new to science; we've known about them for only around

130 years. Vaccines against viral infections were developed by Louis Pasteur in the late 1700s and Edward Jenner in the mid-1800s, but neither really knew how these vaccines worked or what they worked against. Pasteur looked for but couldn't see the pathogen, correctly speculating that it was too small to be seen under a microscope. Scientists in 1892 then began experimenting using filters that would strain out and remove bacterial cells and anything larger, discovering that the remaining filtrate was still infectious. Filtrates of the tobacco mosaic disease, foot-and-mouth disease and the mosquito-transmitted yellow fever could all be used to infect an unlucky and previously healthy new host. Finally, in 1928, the virologist Thomas Rivers concluded from his research that 'viruses appear to be obligate parasites in the sense that their reproduction is dependent on living cells'.[4] At that time, viruses were thought to be amongst the simplest living organisms.

In 1935, viruses received a demotion to not being alive after all. Instead they were again considered poisons or biological chemicals. This demotion occurred after a group at Rockefeller University crystallised a virus and found that it consisted of only a package of complex biochemicals. Later work by these and other research groups around the world established that viruses consist of a genome based on nucleic acids (DNA or RNA) which are enclosed in a protein coat. As virologist Luis Villarreal has noted, 'By that description, a virus seems more like a chemistry set than an organism.'[5]

Viruses typically have very small genomes, probably because they borrow so much of their host cell's machinery to reproduce. RNA and DNA are both made up of four simple coding molecules called nucleotides: cytosine (C), guanine (G), adenine (A) or thymine (T). Viruses, bacteria, bees, birds and all life forms have DNA or RNA genomes that are composed of these four nucleotides that are arranged in long strands. The cellular machinery converts these codes into proteins or other molecules within the cell, and it is that machinery that is co-opted by the viruses to make more viruses. Villarreal describes the process:

When a virus enters a cell (called a host after infection), it is far from inactive. It sheds its coat, bares its genes and induces the cell's own replication machinery to reproduce the intruder's DNA or RNA and manufacture more viral protein based on the instructions in the viral nucleic acid. The newly created viral bits assemble and, voilà, more virus arises, which also may infect other cells.[6]

The entire DNA or RNA genome of viruses is tiny. The Kashmir bee virus (KBV), for example, has a genome consisting of a little over 9500 base pairs. I've included in this chapter a figure of the near-complete KBV genome, printed on just one page of this book. That genome is a sequence we extracted from a KBV-infected colony of Argentine ants here in New Zealand.[7] This 'bee' virus is also common in invasive wasps, ants and many other insects. In comparison with the 9500 base pair size of KBV genome, the genome that you carry in every

Pictured is the near-entire genome of the Kashmir bee virus, which colleagues and I sequenced from an infected sample of Argentine ants. One of the reasons viruses have comparatively small genomes is their ability to co-opt their host cell's machinery.

one of your body cells consists of 3.23 billion base pairs. If I were to print that genome as I did for KBV, it would take up 340,508 pages. That book would be very boring for you to read, but it is perhaps the greatest treasure scientists have for advancing human medicine and health.

Are viruses actually alive? I tell my first-year biology students that a living organism has a range of features. It has an ability to regulate its internal environment, to grow and reproduce, to transform energy by converting chemicals and energy into cellular components, and to evolve. Rocks have none of these characteristics, while viruses have some but not all. Viruses cannot regulate their internal environment, carry out metabolic activities, or reproduce by themselves when outside of the host cell. However, their ability to evolve and change is outstanding. They often have a massive abundance within hosts, a fast generation time, and a lack of the cellular machinery that organisms like you and I need in order to correct any errors when our DNA is copied. Viruses have been estimated to have a mutation rate during reproduction that is nearly a million-fold higher than that of animals.[8] Most of the genetic mutations in those viruses are likely to either have no effect or cause the death of the virus. But some mutants will survive and have traits that are beneficial. Perhaps these mutant genotypes have a way of avoiding the immune system of their host. It is these viral genotypes that might live and preferentially reproduce, damaging living things like bees, and people.

This is a good place to make a quick note on virus systematics and viral taxonomy. The mere mention of taxonomy might be enough to make you think, *Boring*. But taxonomy is anything but dull. Biologists are known to have intensely heated arguments about species definitions. If you can imagine a red-faced nerd jumping up and down at a conference, yelling at a colleague for having the gall to suggest that his favourite plant is indistinguishable from some other plant species that few people care about, you'll have it just about right. Viral taxonomy is often just as controversial among virologists. It's also difficult to understand, and largely based on nucleotide sequences. But there is no hard and fast rule about what degree of variation defines a viral 'species' instead of a 'strain'. Are two viruses that are 83% similar in their nucleotide sequences the same species, two strains of the same species, or two different species? There is no set rule for this sort of determination.

The scientific or Latin name for the western gorilla is, unimaginatively, *Gorilla gorilla*. The western honey bee is *Apis mellifera*. The name is the genus followed by the species. Latin binomial names of genus and species, however, aren't commonly used with viruses. For example, the *Deformed wing virus* is the actual species to which the virus belongs. It is the same with the *Israeli acute paralysis virus*, the *Kashmir bee virus*, and the *Acute bee paralysis virus* – each of these is the 'species' name (though, as we will see, there is debate about whether these three should be divided up as individual species or lumped together as one species and three strains). When written without italics, each is also the name of the virus itself. A virus might be named after the area or geographic features near its discovery, the symptoms associated with infection, or the host from which it was first discovered.[9]

At higher levels of classification there is more consensus. Viruses are typically classified by the Baltimore classification system, whereby they are placed in one of seven groups. These seven groups depend on a combination of their nucleic acid (viruses can have genomes based on DNA or RNA), strandedness (where their genome is single-stranded or double-stranded DNA or RNA), 'sense' (how their nucleotides are translated into proteins), and the method of viral replication. The deformed wing virus is in Group IV: it is a positive-sense single-stranded RNA virus, which is coded as belonging to the (+)ssRNA group.

Viruses lead a 'borrowed life' that most biologists would define as being in the grey area between life and chemicals. While viruses cannot really be considered 'alive', in every practical sense for your honey bees they might as well be. Scientists might jump up and down and argue about viral classification and definitions and whether they are actually alive – but, for the typical beekeeper, this package of complex viral chemicals is very tricky. Frequently, it acts like a living pathogenic organism that devastates honey bees.

Covert and overt 'mutant clouds' of viruses in honey bees

Up until 2017, only 24 different viruses were known to appear in honey bees.[10] That's an amazingly small number and we now know there are many more, which I'll get to later.

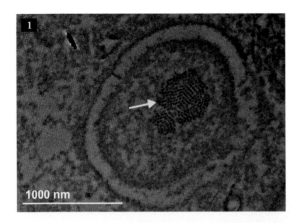

1. An electron microscope photograph shows virus particles beautifully arranged in rows in a cell of the common wasp *Vespula vulgaris*.

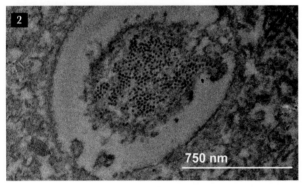

2. A cell is bursting open, releasing virus particles that will infect other cells within the wasp.
Photos: Oliver Quinn

One big reason so few of these pathogens were identified was that only the viruses that are overtly expressed were easily seen and identified. If a honey bee pupae has a high DWV infection, they typically have shrivelled, deformed, useless wings as an adult. Similarly, juvenile bees with substantial sacbrood virus (SBV) infections have their skins harden, coming to resemble a dead, fluid-filled sac. These infections are 'overt'.

While we've known about these pathogenic viruses for many decades, what wasn't understood was how many strains or variants of these viruses there are. A single honey bee host might pick up different strain infections of the same virus from different sources, or even contain a 'cloud of mutant' viruses that have diverged from their original infection.[11] Different strains or mutants can have different effects on their hosts.

The covert viruses, those that have few or no obvious effects on their hosts, have been harder to spot. They have been given odd names like *Apis mellifera bunyavirus* 1 (ABV-1). But these viruses still may have a role to play in honey bee colony losses. We know very little about these viruses and their influence on honey bees. Perhaps they have their greatest influence on bees in conjunction with other pathogens, or possibly even in the presence of pesticides. Perhaps they have no influence on bee health at all. Perhaps some of them are actually beneficial to their hosts.[12]

I'll focus on the viruses that have an overt and substantial effect on honey bees. Of these pathogens, DWV is easily the most devastating. But first, how do bees become infected?

How are viruses transmitted between bees and beehives?

In the previous chapter I talked about that parasitic mite *Varroa destructor*. By feeding on bees, *Varroa* transmits and spreads a wide range of viruses, such as DWV and other pathogens. *Varroa* is like a dirty hypodermic needle, though with viruses replicating or breeding within the mite prior to being injected into a new bee host. *Varroa* has distinctive strains of viruses that live in the mite and that parasitise bees. Scientists have been able to map the source and spread of DWV by following global colonisation of *Varroa*.[13]

There are many other ways viruses can spread among bees and beehives. Adult bees in the hive clean cells that contain faecal deposits, then remove diseased or infected larvae and adult bees from the hive, typically using their mandibles and mouths. These bees then use their mouths to move food around the hive and to the brood bees. Close contact between disease and food is a sure pathway for the spread of disease.

When infected insects forage on flowers, they can also contaminate this food source with viruses from their faeces or oral saliva. Subsequent foragers can then acquire an infection. It's possible that adult bees won't become infected from the low levels of viruses collected from flowers, but the juvenile bees that the pollen is fed to are more susceptible to these pathogens.

Because many different insects host bee viruses, viruses can also be passed between various bees and pollinating species. In England, for example, a recent study showed a high correlation between infection by DWV in bumble bees and honey bees.[14] These pollinators only interact on flowers, so it seems likely that sharing the same food source is a major method of viral transmission. There is also evidence that *Varroa* mites themselves can be deposited on flowers. The patiently waiting mites will then leap nimbly on board the next honey bee forager to come along.[15] These honey bee foragers acquire a little bit of pollen or nectar, a *Varroa* mite, and the package of viruses and pathogens within the mite. Flowers are dangerous places.

Drifting and robbing behaviours can also spread viruses. Drifting behaviour is especially common when hives are located close together. Bees sometimes seem to genuinely mistake their home hive for another, with the guard bees allowing their passage. Those bees will work in the hive, foraging and sharing whatever pathogens they have with their new home. Perhaps the effects on learning and memory after infection by DWV facilitate drifting and, subsequently, the spread of the virus. Honey bees also rob other hives of their honey, especially weaker hives that may be infected by viruses. Viruses present in the honey are then transported back to the new hive. *Varroa* mites will hitch a ride then, too.

The deformed wing virus

Scientists think that the deformed wing virus (DWV) 'is now the most likely candidate responsible for the majority of the colony losses that have occurred across the world during the past 50 years', as written in a 2012 *Virulence* study.[16] This virus has caused the collapse of millions of honey bee hives around the world due to its mutualistic relationship and close association with the parasitic mite *Varroa*.

DWV was first described as being from Egypt in 1977, so at that stage it was called the Egypt bee virus. The bees from which it was described didn't have deformed wings, and it wasn't until 1982 that a group in Japan discovered the virus in beehives infected with *Varroa*. These bees did have deformed wings, and thus a new name for the virus was born. Although *Varroa* can vector many viruses and other pathogens, it is DWV that is typically the most prevalent in *Varroa*-parasitised bees.

What are the symptoms of an overt, virulent infection of DWV? It is accepted that this virus is indirectly or directly responsible for the deformed wings of emerging bees.[17] The damaged wings are particularly stubby, shrivelled and useless. The bees cannot fly. Overtly infected honey bees also have shortened, rounded abdomens. Their legs and wings are discoloured and they are often less hairy than healthy bees. Symptomatic bees have a severely reduced lifespan (less than 48 hours, usually) and are typically expelled from the hive by their sisters. Bees that are less damaged or infected are still able to fly, and they start foraging at an earlier stage, putting additional stress on the colony.

Less obvious is a reduced ability to learn. Bees become less able to form memories and to react to stimuli such as sugar concentrations in nectar or liquid.[18] Also less obvious are the *Varroa*-driven effects associated with the beneficial bacteria and other pathogens in bees. It is difficult to distinguish the effects of DWV from the effects of *Varroa*. One of the many effects of *Varroa* parasitism is a change in the bacteriome of the bees, meaning that they have lower levels of beneficial bacteria. In a 2016 study, my colleagues and I found that, of all the pathogens measured, it was *Varroa* (and the viral pathogens it carries) that had the largest effect on the bacterial community of the bees.[19] Some of these bacteria were pathogens, while others were essential to the functioning of the colony.

In the absence of *Varroa*, DWV is thought of as, basically, a puppy dog. Though your bees might have a low abundance of multiple strains of the virus, these strains form a covert infection that doesn't result in any visible symptoms or any apparent negative impact. It's thought that these infections are persistent over the lifespan of insects and indicate that DWV is unusually well adapted to its honey bee host.[20] There is, however, some emerging evidence that the covert, hidden infections of this virus can indeed have negative effects. One study found that adult workers injected with low doses of DWV started foraging at an early age and that the virus reduced their total activity span as foragers.[21] That study, however, still injected the virus in the way used by *Varroa* mites. No significant 'covert' effects of a wide variety of different strains of DWV were observed in

Hawai'i prior to *Varroa* introduction: it was only after the arrival of *Varroa* that colonies began to collapse. Our best guess is that DWV needs *Varroa* (or an injection) to become virulent and nasty for bees.

There are many different variants of this virus. Before *Varroa* was introduced into Hawai'i, at least 15 different variants or genotypes were observed.[22] But after *Varroa* was introduced, just one variant dominated the bees. Scientists refer to this *Varroa*-vectored strain as deformed wing virus B (or DWV-B; previously known as the *Varroa destructor* virus-1). This strain appears much more overt and pathogenic than the other main strain recognised as deformed wing virus A. An increasing prevalence of the nasty DWV-B has been closely linked with overwinter honey bee worker loss.[23]

We humans have spread DWV around the world, because as we have exported bees we have also unintentionally moved *Varroa* and this virus. It is a little ironic to me that the international export of other viruses, such the virus that causes foot-and-mouth disease in livestock, would cause a massive uproar and market access problems. Yet the international community appears relatively

Deformed wing virus is common in the honey bee. It has also been observed in species such as the buff-tailed bumble bee, and wasps and hornets. *Photo: Phil Lester*

unbothered by exports that contain 'the most likely candidate responsible for the majority of the colony losses that have occurred across the world during the past 50 years'.[24] The global exchange and international export of bees is still happening. New Zealand is a major exporter of bees (and bee viruses, probably with the occasional *Varroa* mite) to Canada. Others have recognised this export is a major issue, such as Lena Wilfert, who has documented the global movement of *Varroa* and viruses. 'People didn't on purpose do this,' she told the *Washington Post* in 2016. 'People don't go to the trouble of sending bee queens to the States for stupid reasons. They do it to get better hives or honey, to get more pollination. Until recently we didn't understand how common it is to spread diseases that way.' Trade should be thoroughly policed at national ports to keep the problem from worsening, she argued. 'If there are any mites around, there's no question they shouldn't be traded at all.'[25]

Bizarrely, Australia does not currently have DWV (although there are one or two researchers who believe that they have seen it there).[26] If or perhaps when *Varroa destructor* establishes itself in Australia, it will be related to human trade. A bee swarm might arrive surreptitiously on a ship, or a beekeeper might bring some bees along with him in his pocket. And when *Varroa* arrives in Australia, the country will be introduced to the special strain of DWV that *Varroa* carries too. I've searched for DWV in Australian bees and other insects, as have others, and don't understand why DWV is at least largely absent: Hawai'i had this virus prior to the introduction of *Varroa*. Why isn't it in Australia too? Perhaps the biggest concern for Australia might be the effects of DWV on native bees and insects that are evolutionarily naïve to this virus. There are 1600 native bee species in Australia that have never experienced DWV. How will they cope when it sweeps through pollinator communities, as it has elsewhere?

Sacbrood virus and black queen cell virus

The deformed wing virus is nasty and has caused widespread devastation to honey bee colonies. But some scientists believe there are much worse viruses, and they're common to most countries. Two of them are the sacbrood virus (SBV) and the black queen cell virus (BQCV). Inadvertently, it seems that the *Varroa* mite selects against highly virulent and lethal viruses such as SBV and BQCV. A team led by Emily Remnant and Madeleine Beekman of the University of Sydney carried out an experiment in which they injected several generations of bees with SBV, BQCV and DWV. Injecting bees with these viruses is the equivalent of feeding by *Varroa*. After injection, both SBV and BQCV out-competed DWV. High rates of bee mortality were observed.

The authors concluded that if *Varroa* carried viruses that kill honey bees quickly, the mite wouldn't be able to complete its development and produce new generations of mites. Over successive generations of *Varroa*, it seems that these lethal viruses are eliminated from the mites, leaving DWV as the primary viral pathogen. This virus doesn't immediately kill the honey bees, and consequently the authors think that we shouldn't be overly concerned about DWV.[27] Beekman suggests that this 'dangerous' bee virus might even be just an innocent bystander: 'If we want to protect the bees, it now no longer seems to make sense to try to combat the virus. Instead, there needs to be a renewed focus on ensuring

A honey bee queen cell, showing a discoloured queen bee larvae infected with BQCV. This virus causes mortality in queen bee pupae. *Photo: James Withington*

the number of mites in honey bee colonies remain low.'[28] Personally, I'm not convinced that we shouldn't try to limit the effects of *Varroa* by combating its mutualistic DWV, but I think this study does nicely demonstrate how this parasitic mite has probably developed a relationship with a virus that is typically less lethal than others.

BQCV has been considered one of the most common and prevalent virus infections in honey bees worldwide.[29] I'm hoping that you can guess what the virus is primarily known for: the queen pupae and pre-pupae are found dead, black and decomposing in their cells. BQCV is considered to be the most common cause of death in queen bees. Infection of worker bees also occurs. Hives that are weakening can have high infection levels of this virus, in addition to a suite of other viruses and pathogens. Just as for DWV, BQCV can be found in other insects, such as bumble bees and other pollinators. It's widespread, and a common infection of invasive Argentine ants in New Zealand.[30]

SBV is another common infection in honey bees, though it is seldom observed in other insects. Infected workers typically have covert infections – they seem healthy and appear not to be suffering any effects of the virus. Larvae and juveniles are infected when an adult nurse feeds them virus-contaminated larval food. Highly infected larvae form a sac-like body and fail to pupate, dying in their cells. Workers are likely to become infected when cleaning out the cells with larvae killed by this pathogen. Although SBV is frequently lethal for bee brood, hives that have this infection seldom collapse. Consequently, SBV is not considered a significant threat to managed honey bees. This virus has been detected in *Varroa*, though it doesn't reproduce or replicate in this mite, which might suggest that the mite doesn't serve as a common host or vector for this pathogen. Perhaps Emily Remnant and Madeleine Beekman are right: *Varroa* might just keep clear of the really lethal, nasty viruses.

The AKI-complex: Three quasi-species of virus

'The AKI-complex' sounds like a *New York Times* bestselling thriller, doesn't it? Unfortunately, this complex is not as exciting, and it is bad news for bees.

Many beekeepers will have heard of these three viruses: the acute bee paralysis virus (ABPV), the Kashmir bee virus (KBV) and the Israeli acute paralysis virus (IAPV). These three 'species' have been known and named from bees for decades, but only recently have viral taxonomists decided that they are very closely related and probably not distinct species. We now think of them as quasi-species, part of a 'mutant cloud' of viruses that are closely related.[31]

AKI-complex viruses can be lethal and highly virulent. With both ABPV and IAPV, infected adults show a rapidly progressing paralysis. They tremble, cannot fly, and show a gradual darkening in body colour and a loss of hair from the abdomen and thorax, before dying. Comparatively, there are very few described symptoms for KBV: bees infected with this virus just seem to die. The most obvious sign of a severe infection of KBV is a sharp decline in a bee population.

KBV was first identified in 1977 in adults of the Asian bee (*Apis cerana*) from the northern and western regions of India. The researchers at that time attributed the Asian honey bee as the likely original host and observed that bees died within six days of being injected with this virus, although western honey bees could eat food contaminated with KBV and appear unharmed.[32] This result, whereby adult and pupal bees that are orally infected with this virus become covertly infected carriers with no apparent ill effects on their health, has been repeated by researchers again and again. We've also seen repeatedly that if you inject KBV into honey bees, the virus quickly replicates and kills the bee within just a few days.

These results suggest that it would be devastating for bee colonies if a parasite such as *Varroa* became infected with KBV and inadvertently injected it while feeding. KBV was implicated as just such a link between the deaths of honey bee colonies in New Zealand after the arrival of *Varroa* in 2000.[33] Many honey bee hives with a lot of *Varroa* were also highly infected with KBV. Similarly, this virus has been associated with colonies displaying colony collapse disorder symptoms in the United States.

Colonies observed to be collapsing will often have high levels of one, two or all three of the AKI-complex viruses. For a while, colony collapse disorder was highly correlated with IAPV and it was thought that this virus was a likely culprit.[34] Scientists believed that this virus could be a major player in colony collapse, alone or in concert with other players. And, when hives are infected, they often experience a massive infection. Viruses such as IAPV infect or contaminate pretty much everything in the hive. A 2014 study led by Yan Ping Chen found this virus in 'eggs, larvae, pupae, adult workers, drones, and queens as well as *Varroa destructor* that fed on the bees'. Every bee body organ examined had this virus present. In addition, it was detected in royal jelly, honey, pollen, queen faeces and drone semen. It was the third most common virus present in the bees after DWV and BQCV.[35]

Chen's work suggests that the function of bee mitochondria is strongly affected by IAPV. Mitochondria are the powerhouses of cells for me, you and your honey bees. An IAPV infection will consequently cause a disturbance in energy-related processes within bees and beehives. The symptoms of this virus, including trembling and an inability to fly, sound like an energy-related disturbance. (A *Star Wars* fan would probably make some joke here about a disturbance in the force, but I'm not a *Star Wars* fan.)

The studies suggesting that one or more AKI-complex viruses are implicated in colony collapse are typically based on correlations. Researchers will sample a weak hive that appears to be near collapse and will find high levels of *Varroa*, viruses and other pathogens. The viruses may be a secondary infection after the bees' immune systems have been weakened by *Varroa* mites feeding, or perhaps the viruses have allowed the mites to attain high densities. It's also possible that the mites and viruses have become abundant when the bees were stressed by factors like poor nutrition or exposure to a nasty pesticide. The effects on bees of *Varroa*, viruses, other pathogens, and the wider environment are complex and compounding, meaning that it has often been difficult to isolate the cause of a colony's collapse. Typically, though, *Varroa* is involved.

Just as for DWV, none of the three viruses in the AKI-complex can be found everywhere. For example, IAPV is prevalent in Australia but New Zealand doesn't seem to have it at all. A study published in 2014 set out to undertake a thorough analysis, with 1050 tests carried out from 499 apiaries across the country, and no IAPV was found.[36] The absence of this virus has considerable implications for countries that would like to import honey into New Zealand: the government can legitimately exclude imports from everywhere else on the basis of the risk of contamination with IAPV. To exclude IAPV and other pathogens, it is also illegal to import bees, bee semen and even pollen into New Zealand.

Other viruses found in honey bees

The viruses I've described above are the ones we know the most about, primarily because of their widespread nature and their periodically devastating effects on honey bees. As mentioned earlier, until very recently only 24 different viruses were thought to infect honey bees − which isn't many, considering you might find 10 million viruses in one drop of seawater.[37] Nowadays, genetic tools, high-performance computers and some nifty software allow researchers to efficiently hunt for viruses. A single study in 2017 added another seven viruses to the list of 24.[38] They were given names like *Apis mellifera* rhabdovirus 1 and *Apis mellifera* bunyavirus 1. Some were widespread and associated with *Varroa* and the fungal pathogen *Nosema apis*, and the bees in which they were found were nearly always infected with DWV, SBV and BQCV. Others of the new viruses were found at only one sampling site. I'm confident that this list will grow to be in the hundreds − that over the next decade we will see a wide diversity of additional viruses being described from honey bees.

On the list of 24 viruses that we've known about for some decades there are some familiar foes, such as the aphid lethal paralysis virus, the Big Sioux River virus, the chronic bee paralysis virus, the Lake Sinai virus, and the slow bee paralysis virus. Many of these have a wide range of arthropod hosts. Some can also live inside of, replicate in, and be vectored to honey bees by *Varroa*. And sometimes people find high abundances of these viruses in hives that are collapsing. But, just as with other diseases, a high abundance of these viruses in collapsing beehives might be a symptom rather than the cause. It might − or might not − be a correlated symptom of stressed bees with compromised immune systems. As we've seen, a collapsing hive presents a complicated web of interacting pathogens in a complex environment.[39]

Bees heavily infected by the chronic bee paralysis virus (CBPV), for example, are unable to fly and have hair loss. They tremble and clearly have impaired balance and coordination, which is due to damage to their brains and nerves. Historically it hasn't been a common disease, but it can be observed around the world and is occasionally associated with hive collapse. CBPV is known by a variety of names, including 'the hairless black syndrome', with infected bees known as 'little blacks'. These sick bees can have partially spread dislocated wings and bloated black abdomens. The bloating is associated with a form of dysentery. These individuals die within a few days. Severely affected hives can suddenly collapse, often at the height of summer. Usually, all that remains is the queen and a few workers. A recent study in England and Wales has found CBPV to be a serious emerging disease, with an exponential increase in cases between 2007 and 2017. Similar to our global problems with DWV, the researchers found evidence that the increasing prevalence of CBPV was associated with increasing rates of international movement of honey bees around the world.[40]

The symptoms of CBPV have been described as identical to those seen in bees with the Isle of Wight disease.[41] Perhaps the Isle of Wight bees were afflicted by CBPV? Other researchers and historians, however, consider the massive loss of beehives on the Isle of Wight between 1905 and 1919 to be due to a parasitic mite called *Acarapis woodi*, which infects and feeds within the breathing system of bees, or to the fungal pathogen *Nosema apis*, or to environmental conditions that included 'over-enthusiastic beekeepers'.[42] I imagine that the loss of bees over this period was due to some combination of these factors. But we have a tendency to oversimplify and hold some individual cause responsible, as highlighted in the concluding statements of a 1964 review:

> Whatever the causes of their bees' misfortunes, however, beekeepers used Acarapis woodi after it was discovered as the scapegoat and so it inherited the aura of the IOW [Isle of Wight] disease. The significance of the myth is the lack of sufficient knowledge allowed it to develop and dominate thought and cause so much unnecessary apprehension and wasted effort. The moral, I trust, is obvious.[43]

Viral infections compounded by pesticides and other stressors

It is clear that sub-lethal stressors such as pesticides can increase viral infections in bees. I'll discuss the influence of pesticides on viral loads in honey bees later, but here we need to acknowledge the interaction between pesticides and viruses. For example, the effects on bees of neonicotinoids, which are some of the most commonly used pesticides in the world, have been extensively studied. Researchers have observed that increasing doses of neonicotinoids correlate with increasing infections of DWV.[44] Specific pathways within a honey bee's immune system seem to be weakened by the use of these chemicals,[45] so an increased susceptibility to a range of viruses and other pathogens is expected after exposure to neonicotinoids. Similarly, the use of pesticides within a hive to control *Varroa* can cause bee mortality, especially when bees have high virus loads. But not controlling *Varroa* typically results in high levels of bee mortality, and potentially the collapse of the hive.

A bee's immune response to viruses can be similarly weakened by a poor diet. As the authors of a 2015 review point out, 'There is a relationship between the effectiveness of social and individual immunity and the nutritional state of the colony.'[46] Pollen and the protein it supplies is essential for the function of the immune system of bees, as well as influencing longevity and helping with pesticide detoxification. Bees with a protein-deficient diet have been observed to have a higher number of DWV infections.[47] Recent work has shown evidence that high-quality, poly-floral diets can reduce honey bee mortality due to infection with IAPV. This work suggests that 'good' diets, with a diverse array of high-quality pollen, can help bees tolerate virus infections. The authors concluded that 'good nutrition is an extremely important component of bee health, as high-quality diets appear to buffer bees from other forms of stress'.[48] The study found that micronutrient concentrations of iron and calcium in the pollen may also be important to bee health. Programmes encouraging the planting and maintenance of a wide variety of pollen sources for bees, such as Trees for Bees in New Zealand,[49] can only be beneficial.

Virus control now and in the future

I think that *Varroa* control is fundamental for virus control. I really do seem to like finishing chapters with a quote, so I'll do it here again regarding viruses and *Varroa*. The below is from Alexander McMenamin and Elke Genersch in their 2015 review of honey bee colony losses and associated viruses:

> It is without question that *Varroa* infestation poses the most serious threat to the western honey bee colonies and that this is related to the mite's ability to vector virus infections or to exacerbate preexisting infections. Virus infections of honey bees became a serious health problem for entire colonies only after *Varroa* started to infest honey bee colonies. *Varroa* theoretically can take up any virus present in bee hemolymph and, hence, can mechanically vector any virus back into the hemolymph when feeding on the next bee or pupa. Therefore, it is not surprising that since the introduction of *Varroa* the prevalence of honey bee viruses and infections increased. However, for most viruses, conclusive evidence linking them to colony losses is still lacking despite the nearly ubiquitous presence of the virus vector *Varroa* in the honey bee population. Only members of the ABPV clade [described in this chapter as the AKI-complex] and the DWV clade are thought to play a major role in colony losses in the presence of mite infestation.[50]

If we can control *Varroa*, it seems we will be able to limit the influence of viruses, which together are a major cause of global honey bee colony losses. Pesticides, pollen availability and other factors contribute to honey bee viruses too. All of those factors are worth future research for the health of these important pollinators; however, to limit the negative influence of viruses on honey bees, *Varroa* control may well need to be our primary focus.

There might be other treatments that will help combat viruses in honey bees. We can't directly vaccinate bees against them, because bees don't have immune systems that allow vaccines to create a 'memory' that primes them against future infections (although a potential pseudo-vaccine against pathogens such as American foulbrood is being developed, as I'll discuss in the next chapter).

Other recent research has come close to a vaccine, or at least some sort of antiviral drug, using magic mushrooms. Well-known American mycologist Paul Stamets found that certain fungi appear to have antiviral potential for honey bees. He led a study examining mycelia (white filaments of fungi, frequently found in the soil) from amadou and reishi mushrooms.[51] Stamets's study was based on an observation that bees were feeding from water droplets on mushrooms in his garden. His team created a broth from the mycelia and fed it to the bees. Bees that consumed the broth had a 79-fold decrease in DWV in less than two weeks. A 45,000-fold reduction was observed in another viral species, the *Lake Sinai virus*. Stamets said, in a 2018 interview, 'These aren't really antiviral drugs. We think they are supporting the immune system to allow natural immunity to be strong enough to reduce the viruses.'[52]

Providing antiviral mushroom broths, or key pollen sources, could help bees to help themselves.

4. AMERICAN FOULBROOD

My hive smells fishy – now I'm obliged to burn my bees

I've seen beekeepers in tears after discovering American foulbrood (AFB). They are usually amateur beekeepers who have a single apiary site stocked with a small handful of hives. The bees are their treasured friends and pets. The law in New Zealand states that you are legally obliged to kill and burn your bees, along with their hives and honey, once clinical symptoms of this bacterial disease are discovered. After you seal and pour petrol into a hive there is an immense buzz from the bees. And soon there is silence. Some will tell you it is like losing a member of their family.

We first see references to what was probably AFB over 2000 years ago, in the writings of Aristotle. He described disease-like symptoms in honey bees, which 'is indicated in a lassitude on the part of the bees and in malodorousness of the hive'.[1] The name 'foulbrood' was first used in 1766 by the priest Adam Gottlob Schirach. One of Schirach's more controversial – but ultimately correct – observations was that all working bees, rather than being sexless, were 'females in disguise'.[2] Schirach also observed that if a queen bee goes missing from the hive, the workers enlarge cells containing young worker brood, which are then fed a diet that results in the production of queens. This idea was also greeted with hostility and scepticism, because the prevailing thinking at this time was that rulers have a natural hierarchy. Queens and rulers were thought to descend from a great lineage, so royalty could not arise from some random youngster, irrespective of what or how they were fed.[3]

Much less controversially, Schirach described two diseases characterised by fouling larvae: the 'false'p and the 'real' foulbrood pest.[4] It is likely that the 'false' was what we now call European foulbrood (caused by the bacterium *Melissococcus plutonius*, which I'll discuss in Chapter 7) and the 'real' is what we now know as American foulbrood. AFB was formally described and given its first genus and species name, *Bacillus larvae*, in 1906. Almost a century later, AFB received a reclassification of genus and species, to *Paenibacillus larvae*.[5]

Before the arrival and devastating effects of the parasitic mite *Varroa destructor*, the bacterial diseases causing American foulbrood and European foulbrood were the most economically important diseases of honey bees worldwide. As the microbiologist Elke Genersch has written, AFB 'is still among the most deleterious bee diseases'.[6] This globally abundant disease affects amateur and professional beekeepers alike. As we will see, it is an 'obligate killer'.[7] It is one of the few honey bee diseases that can cause the complete collapse of the infected colony, killing individual larva and potentially the entire hive. Beekeepers in many countries, however, are required by law to find and destroy the bees before AFB has the opportunity to exert such a high mortality.

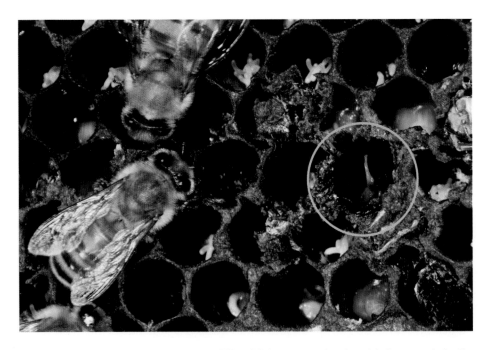

Circled is a larva that has died from AFB, with its tongue (proboscis) characteristically extended. This hive has other problems, including having no queen, as evidenced by the multiple worker-laid eggs in the other cells. *Photo: Phil Lester*

Endospores: Infectious little AFB seeds fed to larvae

Given the name 'American foulbrood', you might have guessed which life stage is affected by this pathogen. Bacterial spores, or more specifically 'endospores', are infectious to only the larval stages or the brood. Endospores lead us to one of the key issues for the management of this disease.

A spore is a reproductive unit that is often adapted for dispersal and for survival. A spore can generally survive unfavourable conditions for extended periods of time. Endospores are an even more remarkable type of spore, produced only by some species in the bacterial phylum Firmicutes. This phylum is named for its members having a strong skin – the endospores of AFB have a tough, dormant and non-reproductive structure. Endospores are simply amazing. These little seeds can survive for a *very* long time. In 1995 Raúl Cano and Monica Borucki of California Polytechnic State University identified, revived and cultured bacterial endospores from the abdominal contents of an extinct bee species.[8] This bee had been entombed in Dominican amber between 25 and 40 million years. The *Jurassic Park* movies are based on a similar scenario, where dinosaurs are revived from blood in mosquitoes preserved in amber, which could never happen (the DNA just wouldn't be preserved for that long). But here we have the actual resurrection of an extinct bacterial species from endospores that were sleepily dormant for tens of millions of years. These and other endospores survive without nutrients and are resistant to pressure, high temperatures, freezing, disinfectants and radiation. At the risk of labouring a point, the tenacity of endospores is amazing.

We don't know if AFB endospores could survive for millions of years. The best research we have, however, does indicate that they are hardy and long-lived. Ongoing work has demonstrated that these spores survive for more than 40 years.[9] AFB endospores can withstand cold and hot temperatures, humidity and drought.[10] Given the ability of related bacteria to survive tens of millions of years, and that AFB has so far been seen to survive at least decades, I strongly suspect that AFB endospores can cause viable infections after lying dormant for centuries. The period of 'centuries' probably isn't relevant to our current management issues with AFB in countries like New Zealand. But very relevant to AFB management is the fact that these endospores can survive for several years or decades on disused equipment or in old feral honey bee nest sites.

When an individual larval bee infected with AFB dies, its body effectively becomes a pool of nutrients for the bacteria. The larval cadaver, as in textbook descriptions of AFB, turns into a sticky, ropy, soupy mass of bacterial food. What results is the production of millions, even billions, of endospores from that single decomposing bee. All it takes is a tiny number of endospores for new larvae to become fatally infected. It's been estimated that approximately 50% of newly hatched larvae will die if given just eight endospores.[11] These young honey bee larvae are very susceptible to AFB infections. As far back as 1966, scientists were finding that larvae less than six hours old needed only a few spores to become infected. At 18–24 hours old the larvae needed many more for a successful infection, and 36–42-hour-old larvae had developed resistance to infection at any experimental dose attempted.[12]

After ingestion, the endospores 'hatch' and proliferate in the larval gut. The

A larva is usually infected by nurse bees placing contaminated food in its cell. The larva eats the infected food and spores. The spores germinate in its gut, and multiply. A dead larva such as this can contain billions of endospores. The cycle of infection begins again when a nurse bee cleans out the cell and feeds other larvae. *Photo: Phil Lester*

bee gut has two major layers. The bacteria produce an enzyme that eats into and degrades the first of these layers – the peritrophic matrix. Toxins or other unknown factors, depending on the AFB strain, then allow the bacteria to penetrate the other layer – the epithelial cells.[13] Eventually, the combined effect of the enzymes and toxins allows the bacteria to breach the gut and enter the body cavity of the larvae. As the larval bee ages, changes in its gut likely make it less suitable for AFB. Perhaps the pH of the gut or the epithelial cells change in some way that makes it difficult for the bacteria to proliferate and penetrate the gut wall.

The larvae acquire infections from adult nurse bees. When an adult worker bee cleans cells that have AFB-killed larvae inside them, she becomes an endospore carrier and spreader. This bee is immune to the effects of AFB, but if she feeds or tends a larva of just the right age, the infection process will begin again. As Elke Genersch writes, the 'death of infected honey bee larvae and total degradation of the larval cadavers are the prerequisites for transmission and spreading of the disease within and between colonies'.[14] The long-lived and tough little endospores allow the bacteria to persist in a hive, an apiary, a region and a country.

How different strains of AFB affect bees

In previous chapters I've talked about how different strains of pathogens such as viruses have different effects on their hosts. The same is true for AFB.

Historically, AFB has been classified into four 'ERIC' strains. ERIC is a molecular classification. If you really want to know what it stands for: Enterobacterial Repetitive Intergenic Consensus. Feel any wiser? The reasons for the genetic differentiation aren't important; the key titbit is that there are different strains. Of the four, ERIC I and ERIC II are the relevant ones. They correspond to the previously used genus, species and subspecies classification of *Paenibacillus larvae larvae*. The other two AFB strains haven't been observed in the field for decades: ERIC III and ERIC IV are reported only in laboratory culture collections. Both fall into an older and currently disused sub-species classification of AFB, specifically the *Paenibacillus larvae* subspecies *pulvifaciens*.[15] A fifth strain, ERIC V, was recently discovered[16] but will likely be debated by the scientific community before broad acceptance.

Some readers might now be thinking, 'Who cares about strains?' Well, beekeepers should care, because the AFB strains kill bees differently. They also cause your hives to collapse in different ways and at different times. Each strain can be present in your hive, killing bees, perhaps for years before you'll see classic and commonly described clinical AFB infection symptoms.

New Zealand beekeepers like myself are encouraged to take a course on the identification of AFB. The textbook description is that AFB infects newly hatched bee larvae. The larvae keep growing and even reach the pupal stage. The cells are capped. Then the AFB-diseased pupae stick their tongues out and die within their enclosed chambers. We are trained to recognise the signs of AFB in these dead bees in capped cells. The cells are typically dark and sunken, often with holes chewed by investigative worker bees. The pupae underneath are coffee coloured. They have their tongues poking out, and they form a ropy, sticky mass like glue when you prod them with a stick. They smell like rotting fish, and the pattern of brood production is 'spotty', without the uniform distribution of larvae across a frame that we see in a healthy hive.

All of these symptoms that we are taught are correct. They nicely describe what is likely a very late-stage AFB infection within a considerably weakened hive. AFB is likely to have been present in this colony for weeks, months or even years prior to the moment the beekeeper or scientist sees dead bees in capped cells. In the earlier stages of infection – perhaps years or months before – the much stronger hive had worker bees cleaning out and removing the diseased or dead larvae.

We've only recently become aware of this disease pattern through work from Genersch's laboratory in Germany. One of the key messages of her work is that larval mortality from an AFB infection isn't limited to a specific time or life stage of the bee. The first larvae to die do so just two to three days after being infected. Further, the majority of infected larvae die prior to the cells being capped. These dead larvae are removed by nurse bees as part of their hygienic behaviour, so they are never seen by the beekeeper. In prior work experimentally infecting and observing patterns of AFB mortality, 'scientists, bee keepers, veterinarians just couldn't and didn't see these larvae and concluded that they did not exist'.[17] It is only in the late stages of infection that we start to see those textbook descriptions of AFB, where the number of diseased larvae is so great that they cannot be cleaned away by the declining number of nurse bees. When using standard visual cues, we only see an AFB infection when colonies are very sick and have been for some time.

This is where differences between the strains of AFB become relevant. In what Genersch describes as the second important message, experimental infections have demonstrated the strains differ substantially in their patterns of virulence (the microbe's ability to infect or damage the host). Each strain, ERIC I and ERIC II, has a different virulence on individual bees. And the consequences of that difference mean that each strain has a different – and quite the opposite – virulence at the hive or colony level. In an ERIC I infection, AFB takes on

Lab testing is necessary for a definitive diagnosis of AFB, but a common field test is to touch a dead larva with a matchstick or twig, or in this case a hive tool. The larva will be sticky and 'ropy' (drawn out). Using a hive tool for this purpose is not recommended, as there is potential to spread spores to other hives.
Photo: Phil Lester

average ~13 days to kill all infected larvae. Approximately 40–60% of larvae will die before the cells are capped. Which means that 40–60% of infected larvae die in capped cells, which then develop into a ropy mass and billions of infective endospores. With such a high number of dead larvae in capped cells, it is comparatively easy to spot an ERIC I infection. And the high abundance of infectious endospores means that more bees will be exposed and the disease will spread quickly through a colony. ERIC I strains often kill a colony quickly.

In contrast, with an ERIC II infection, it takes on average seven days for AFB to kill all infected larvae. Some larvae die sooner than seven days, with a small proportion dying later. That faster rate of mortality results in 80–95% of larvae dying before the cells are capped. These dead larvae are quickly removed by nurse bees. Because of this faster rate of mortality, a smaller proportion of 5–20% of infected larvae die in capped cells. A key consequence is that the build-up within a hive of infective endospores in an ERIC II strain infection is comparatively much slower. The relatively lower and slower accumulation of endospores means that AFB will spread slowly through the hive, with the colony persisting for considerably longer than with an ERIC I infection. ERIC

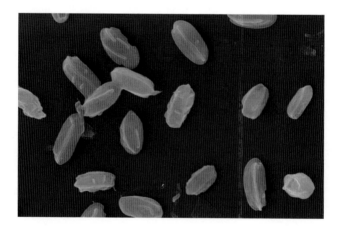

AFB bacterial spores, from the ERIC III strain of the disease. The spores of all strains look similar. Measuring between 1.5µm (0.0015mm) and 6µm (0.006mm), they are extraordinarily tough. They can remain viable for decades in old nest sites or infected equipment, to then become infectious and lethal after ingestion by a larval bee. *Micrograph: Manfred Rohde and Michael Steinert*

II kills your colony more slowly than ERIC I.

I like the way Elke Genersch describes the implications of this difference for the colony's mortality and for disease detection. Below is an excerpt from one of her more recent reviews (bold text indicates key points):

> **Hence, disease development and resulting colony collapse will be much faster in ERIC I-infected colonies than in ERIC II-infected colonies, and *P. larvae* ERIC I must be considered more virulent at colony level than *P. larvae* ERIC II.** The reverse virulence at individual and colony level is counterintuitive, but it is very important to understand this relation: the fast killing and, hence, more virulent genotype at the larval level becomes the less virulent genotype at the colony level, because the bees' social immune response can cope much better with it by effectively removing most (but not all!) of the infected larvae, thereby reducing the pathogen levels in the colony. In contrast, the bees' social immune response less efficiently wards off the genotype *P. larvae* ERIC I that kills larvae much slower and, hence, is less virulent at the larval level; larvae that die too late (i.e., after cell capping) are rather not removed from the colony and instead converted by the pathogen into bacterial biomass and eventually into bacterial spores. Nevertheless, since both genotypes are lethal for individual larvae, they are also lethal for entire colonies and cause AFB outbreaks. However, due to the differences in the numbers of diseased larvae that are not removed by nurse bees and remain in their cells until the final stage of disease, **infections with *P. larvae* ERIC II normally develop much more slowly in the colony and clinical symptoms may be evident only after <u>several years</u> of 'subclinical' disease.** In contrast, colonies infected by *P. larvae* ERIC I develop clinical symptoms much faster and may become obtrusive within a couple of months after infection.
>
> . . . the proportion of AFB-dead larvae that will have died in the capped stage will only be around 10% for ERIC II-infected colonies. In contrast, in ERIC I-infected colonies the same number of AFB-dead larvae will yield about four times more dead larvae in capped cells. **Because only the larval remains (ropy mass or scales) in capped cells are used for the clinical diagnosis of AFB, this diagnosis is four times easier in the case of ERIC I-infections than in the case of ERIC II infections.** Or – in other words – compared to ERIC

I-infected colonies, four times more larvae need to have died in ERIC II-infected colonies before clinical diagnosis is as easy as it is in ERIC I-infected colonies. **In addition, cells cleansed out by nurse bees will be reused for egg laying by the queen. The more cells are cleansed out and again used for larval rearing in the colony, the less patchy the brood nest and the less obvious an ongoing AFB disease will be.**[18]

The two strains are present and have spread around the world. Historically, the ERIC I strain was isolated from AFB diseased colonies on the American continent and in Europe, while ERIC II was thought to have been restricted to Europe.[19] More recently, both genotypes have been found to be widely and broadly distributed around the world.[20] The widespread distribution of the two strains makes sense, given we have been exporting and importing bees around the world for over 150 years. The prevalence of the different genotypes varies substantially. For example, in Japan both strains are present, but ERIC I is the most common genotype.[21] ERIC I and II genotypes are both present in New Zealand,[22] with ERIC II appearing more prevalent in one study examining a small set of samples.[23]

We have known for many years that hives and bee colonies can maintain an AFB infection with low levels of spores for several years without displaying clinical symptoms. Perhaps the colonies that have these long-term sub-clinical AFB infections have the ERIC II strain. Perhaps, also, the aggressive AFB management programmes that require the burning or destruction of hives displaying clinical symptoms mean that ERIC I will be preferentially destroyed, with the more subtle, slowly developing ERIC II becoming more prevalent.[24]

Horizontal spread

Nurse bees infect young individual larvae by feeding them food contaminated with endospores. But how does AFB move between colonies? AFB infections between hives or bee colonies can spread by 'horizontal' (between adult bees) or 'vertical' (from one generation to the next) transmission.

The primary, 'natural' mode of AFB infection has always been considered to be by robbing: a horizontal infection transmission. A late-stage, AFB-infected hive has few guards and a small workforce. Its honey stores are an easy target. Neighbouring bees from a healthy hive determine that another is weak, and they begin thieving on a grand scale. Bees don't seem to notice or care why the hive is weak; in a similar fashion, bees are happy to move into a nest cavity or hive whose occupants have been killed by a disease such as AFB. When robbing bees bring back and store endospore-contaminated honey, spores will be released into the new hive. As the thieving bees move through the AFB-weakened hive, perhaps they also become externally contaminated with spores.

If you place a heavily AFB-infected hive within an apiary, there is a high probability that it will be robbed by bees from nearby hives. The end result is that healthy hives will develop AFB. One study in Finland found that eight of nine hives placed up to 1 kilometre from an AFB-infected hive developed severe clinical symptoms or died from AFB within four years.[25] No hives placed 2 or 3 kilometres from the infected hive died or developed clinical symptoms over the four-year experiment, but these distant colonies were found to acquire AFB endospores at very low levels. This sort of experiment clearly shows that robbing can cause disease in nearby colonies. It's worth noting that in this study, it took three years for AFB symptoms or colony death to occur at a distance of 0.5 kilometres or greater from the infected hive.

Other studies focus on the role of contaminated food as an AFB transmission route. If you feed highly contaminated sugar water to hives of adult bees, the majority will develop an AFB infection. Occasionally, even a sugary cocktail with a light contamination of endospores will initiate the infection.[26] Pollen taken from AFB-infected hives can also contain huge numbers of spores (4.5 million spores per gram), representing another source of potential infection.

As we've already seen, bees drift. Drifting is probably enhanced by beekeepers placing hives close together in apiaries: such a high density of hives in a small area is seldom seen in nature. Bees that drift from one colony to another might spread a disease such as AFB. How often does it happen? The historical consensus is that drifting is rarely successful as a method of horizontal AFB transfer, but I'm not convinced I agree.

In 1994 Mark Goodwin, an apicultural researcher in New Zealand, carried out a key experiment in an attempt to understand the role of drifting in AFB transmission. He and his team used pairs of hives placed together, one with a light AFB infection (defined as <50 larvae showing clinical symptoms) and the other uninfected. They used 25 of these pairings and monitored the bees for an average of three months. Drifting did occur, with just under 6% of marked bees being found in the neighbouring colony after just two days. And AFB did appear to transfer between colonies. Only two (or 8%) of the 25 initially infected colonies developed AFB. The conclusion from this study was that 'drifting of honey bees is not a particularly important cause of the spread of AFB'.[27] That deduction has been echoed by other relatively short-term studies on the role of drifting in AFB transmission.[28]

Goodwin's experiment, however, appears to show that drifting can indeed result in an AFB transfer, even from lightly infected colonies. I also suspect that the AFB transfer rate between hives was much higher than was estimated in Goodwin's work. You'll remember that in the Finnish study on AFB transmission by robbing, it took three years for AFB symptoms or death to occur in hives that were 0.5 kilometres or further away from an infected hive. No hives with AFB symptoms or even deaths were observed within year one, even in previously uninfected hives at the same apiary as the diseased and robbed hive.[29] It was only in year two of the Finnish study that the hives of the thieving bees at the same apiary developed AFB and died. And, in support of the Finnish work, the more recent conclusions are that infections of the ERIC II strain develop much more slowly in colonies, and clinical symptoms may be evident only after several years of subclinical disease presence.[30]

These conclusions on AFB disease development have been reached more recently, and well after Goodwin's study in the early 1990s, in which hives were monitored for a maximum of 388 days (and for an average 103 days). I suspect that if the New Zealand study had monitored hives for several years, more AFB would have been observed and we might now have a very different impression of the role of drifting in AFB transmission.

In the modern age, a key method of horizontal AFB transmission is by the beekeeper. We beekeepers move frames of brood or honey between colonies in order to strengthen a weak hive. We move boxes ('supers') between hives. And we use AFB-contaminated hive tools and equipment on uninfected hives. These beekeeping practices and transmission routes are often considered to have a far greater impact than natural drifting or robbing.[31]

Vertical spread

Honey bees reproduce by swarming. During spring, new queens are produced, and then the old queen leaves the hive with around half the worker force. If this queen and her swarm have come from an AFB-infected hive, will they carry an AFB infection with them?

Unfortunately, yes. Swarms can carry AFB and the disease can spread 'vertically' by this form of reproduction. Many beekeepers have reported captured swarms quickly developing AFB soon after their placement in a hive. Perhaps of most concern regarding AFB vertical transmission is the possibility that sub-clinical levels of the disease are maintained over long periods after swarming events.[32] This additional method of transmission, even if rare, likely provides the pathogen with another foothold and method of persistence and spread in beekeeping operations.

Culling and burning

There are many different methods for the control of AFB, ranging from tolerating and treating the disease to the most extreme method – destroying the hive.

Many countries require beekeepers to destroy any hive that displays clinical symptoms of AFB. Hive destruction by burning is the law in New Zealand, Australia, England, Sweden and Wales. Governments that impose a requirement of destruction often have a goal of country-wide eradication. Hence, no other treatment method of AFB is allowed. You find it, you kill it – end of story. It's not a pleasant task. The standard method involves dousing the hive in petrol, with the fumes then killing the bees. One New Zealand beekeeper described it:

'You hear a really loud buzz, it's like the bees screaming out and crying. It eats at me. It's awful.'[33] Another account of petrol and fire comes from Management Agency, National American Foulbrood Pest Management Plan New Zealand. They cite a group of beekeepers who had more than 100 infected hives to burn. The beekeepers dug a large hole and spent the day dropping into it the hives that they had just killed using petrol. That evening, a match was thrown. The unexpected explosion was heard many kilometres away. AFB frames were blasted for hundreds of metres, propelled by the petrol fumes that had filled the pit.[34] This group learned the hard way that petrol vapours are heavier than air.

The goal of lethal destruction of AFB hives and equipment is often to eliminate the disease from an entire country or region, or at least to limit it to low levels. In New South Wales, AFB is by law the only 'notifiable disease' of bees in an act of parliament. Beekeepers who suspect that it is present in their hives must notify their nearest apiary inspector. Once confirmed as infected, the

A beekeeper burns an infected hive. In many countries, beekeepers are legally required to destroy their hives once they discover American foulbrood.
Photo: Phil Lester

hives and equipment are either burnt or sterilised using gamma irradiation. In both situations, the colony is killed. These methods attempt to minimise, rather than eliminate or eradicate, the incidence of the disease in New South Wales.

How well does the burning and destruction of hives work for AFB elimination or eradication? Let's travel to Jersey, where AFB was first found in 2010, for an excellent epidemiological study.

Jersey is an island measuring 118 square kilometres in the English Channel, about 22 kilometres or 12 nautical miles from the Cotentin Peninsula in Normandy, France. The AFB outbreak here is interesting, because Jersey has a closed, small population of hives with motivated beekeepers who want to control the disease, and because scientists used data from the outbreak to implement 'the first rigorous statistical analysis carried out on a honeybee disease epidemic'.[35]

There were approximately 458 hives in 130 different apiaries on the island. A census and inspection of all hives on Jersey was carried out in June 2010, and in June of each following year. If AFB was present in a hive, then the hive was destroyed and the area scorched in an attempt to guarantee removal of the disease. The offending apiary was re-inspected the following August.

The epidemiology and modelling of the Jersey AFB outbreak unfortunately indicated that 'the epidemic is only very rarely stamped out by the inspection process': only in 3% of 10,000 simulations would epidemics be eradicated using this standard inspection and control protocol. That 3% represents a huge amount of effort for a tiny chance of success. The highest probability of AFB extinction was in simulations in which, after an AFB-infected hive was found and destroyed, every hive in a 3-kilometre radius was checked as well. The maximum chance of eradication using such an intensive approach rose to 45%.[36] I talked to Samik Datta, the lead author of the study, about these results. 'The outcome of 45% is lower than we expected, as within simulations there were often a few hives which were undetected, even with the 3-kilometre radial inspections, which were able to kickstart the epidemic again,' he said. 'The undetected infections may be either "sub-clinical" or merely untested, depending upon the management strategy. A large factor which we could not model due to the absence of data was environmental reservoirs — which apparently can persist for decades. Hence, eradication of AFB may be over-ambitious, compared to something more short-lived outside of a host, such as European foulbrood.'

Computer simulations aside, Jersey authorities initiated a major campaign of destroying hives in 2010. AFB prevalence on Jersey declined over time, until in 2014 just one apiary was found to be infected on the island. Once this infection was eliminated, no additional cases were seen . . . until May 2019. Veterinary officer Theo Knight-Jones said, 'Although we suspected that American foulbrood was still present in Jersey, it had not been detected since 2014. The recent case comes as a timely reminder that beekeepers need to be aware of the disease, and take action to prevent it spreading.'[37] The intensive campaign for eradication on this small, closed population of beekeepers on Jersey appeared to have the substantial benefit of controlling this disease to levels that were effectively undetectable for years. But eradicate AFB it did not.

The Jersey scenario – failed eradication but high level of control – is what scientists such as Elke Genersch might predict. Endospores could remain on some equipment somewhere, or perhaps a subtle infection in one or more hives could bubble away to become 'evident only after several years of "subclinical" disease'.[38]

Living with and treating AFB

A range of countries have put AFB eradication or elimination in the too-hard basket. Some countries have legislation requiring beekeepers to burn hives, in an attempt to regionally control disease prevalence. Others, instead of destroying infected hives, treat or attempt to manage infections. Antibiotics and 'shook swarming' have a long history of use with AFB. Other emerging techniques include immunisation and the use of bacteria-killing viruses called bacteriophages.

Perhaps the oldest method of AFB control is 'shook swarming'. Shook swarming involves taking frames of bees from one hive, shaking them off the frames onto new foundations, then destroying the old combs that have honey, pollen, eggs, larvae and pupae. You effectively start a new colony with the only potential AFB contaminant coming from the spores on or in the adult bees themselves. Any diseased larvae, or the billions of endospores in diseased pupae with their tongues poking out, are destroyed. It is a perfectly legal method for controlling other diseases, such as European foulbrood in the UK. Beekeepers sometimes also use shook swarming for *Varroa* control. Although the legislation of countries like New Zealand do not technically say that shook swarming is illegal, it is not an option for AFB management here. As argued on the internet forum NZ Beekeepers, 'The legislation also doesn't specify that it's illegal to treat AFB hives by dancing round them naked at midnight invoking Gaia's cleansing light or any such twaddle either. It's not stated explicitly at any point, but it doesn't need to be – the practice [shook swarming] is illegal by exclusion from acceptable options.'[39]

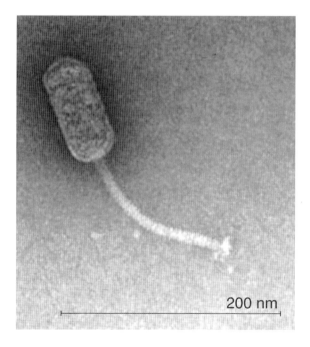

Bacteriophages are viruses that attack and kill bacteria. This transmission electron micrograph shows 'bacteriophage HB10c2', which was isolated from American foulbrood bacteria in bees from Germany. The culture and application of bacteriophages may offer a remedial treatment that controls or limits, rather than eradicates, AFB. *Micrograph: Manfred Rohde and Michael Steinert*

Anti-microbial compounds are used for AFB control in several countries. The broad-spectrum antibiotic drug oxytetracycline has been used in apiculture since the early 1950s. Oxytetracycline and tetracycline-resistant AFB stains, however, have now been detected in the United States, Canada and Argentina. Drug-resistance genes appear to have been transferred between bacterial species.[40] Because of disease resistance and issues around residues in hive products, including honey, antibiotics are banned in several countries. In countries such as France, AFB treatment using antibiotics is acceptable provided the disease is not yet chronic and the honey and wax are destroyed after treatment.

Other anti-microbial or antibiotic compounds have been used to combat AFB. The Chinese government used the broad-spectrum antibiotic chloramphenicol to combat a massive AFB outbreak in 1997–98. Unfortunately, chloramphenicol is a potential carcinogen and has been associated with a disorder called aplastic anaemia in people. Aplastic anaemia isn't especially good for your health: it means the bone marrow is failing to produce enough new blood cells. Consequently, the European Union suspended all imports of honey from China, lifting the suspension only in July 2004. This chemical is still occasionally observed in honey.

Bacteriophages are viruses that infect and replicate within bacteria. Their name combines 'bacteria' and the element '-phage', meaning 'eater'. We've known that bacteriophages can be useful since scientists cured severely induced diarrhoea in calves caused by the common bacteria *E. coli*. These researchers fed the calves bacteriophages they isolated from sewers.[41] In bees, bacteriophages that kill AFB have also been isolated. It's possible to treat and more or less* cure bees infected with AFB, or prophylactically treat bee larvae with bacteriophages, in order to prevent spore infection.[42] In New Zealand, a group at Massey

---

* I've used the 'more or less' qualifier because the bacteriophages are likely to drive bacterial populations to very low levels. But, as with most predator–prey relationships, the 'prey', in this case the AFB bacteria, is unlikely to be driven to extinction. You'll probably end up with the bacteria present at such low densities that they don't cause clinical symptoms or harm to the bees. Bacteriophages are a really exciting area for enhancing the health of bees and people alike. They have been used to quickly (within days) cure people who have been fighting antibiotic-resistant bacterial infections for months.

University are hunting for bacteriophages to use in just such a prophylactic way. 'Our work is not focused on providing a cure for infected hives, but as a measure to prevent infection in a hive before it takes hold,' says microbiologist Heather Hendrickson. 'Bacteriophages have many benefits over antibiotics, as they can be targeted to specific bacteria while the rest of the healthy microbial communities thrive.'[43] Hendrickson told me they have 25 different candidates that beekeepers have provided in the search for an AFB bacteriophage. 'The good news when it comes to bacteriophages is that there are estimated 10^{31} of them on the planet so we have some work to do in finding them, but there is an unlimited supply of the entities in the soil beneath our feet.'

A final developing technique to treat AFB is the use of vaccines. Previously, scientists thought that bees were unable to develop immunity because they lack antibodies, the proteins that help animals to recognise and fight pathogens such as bacteria and viruses. But in 2015 a group from Helsinki found that a protein called vitellogenin binds to pathogenic bacterial species including AFB.[44] Broken segments of the cell walls of the pathogens are incorporated into the vitellogenin. This protein carries these bacterial segments to the eggs, prompting immune responses in the honey bee offspring. Some readers may be imagining tiny needles and syringes – but no, the vaccine isn't injected and instead is given in food.

Bacteriophages and vaccines are both still in development. These techniques will take many years of further work before they become a tool in our arsenal against AFB. At the current time, an array of management techniques can and should be used together to battle AFB. Programmes in Sweden have substantially reduced AFB by the careful and early monitoring of spore levels, shook swarming of bees with subclinical AFB, and the burning of hives displaying clinical symptoms. This 'integrated management' approach substantially reduced disease incidence.[45] It is clear, however, that these approaches, or the use of bacteriophages or antibiotics, are not congruent with eradication or elimination. You can't try to both eradicate and live with the disease. Managers need to choose which approach to take in managing AFB in a country.

New Zealand as a case study for AFB management and potential elimination

American foulbrood was first observed in New Zealand in 1877. It had spread throughout the country and was blamed for a 70% reduction in honey production just 10 years later.[46] The reported rates of AFB peaked at 1.2% of New Zealand hives in 1990. This rate may have been substantially underestimated, however, because in 1993, 12.5% of hives owned by beekeepers who managed fewer than 50 hives were found to be infected with American foulbrood.

AFB is now the single bee disease for which the New Zealand government has legislation to manage. The goal of the AFB National Pest Management Plan is to eliminate American foulbrood in managed colonies (i.e. beehives) in New Zealand by 2030. Mark Goodwin explained the logic and history behind this goal in 2005. First, New Zealand is an island with restrictions on bee imports. Second, AFB is frequently considered not very infective. Third, some commercial beekeepers appear to have eradicated AFB by management, which suggests others can achieve eradication too. Fourth, although the endospores can

American foulbrood detector dogs are part of a burgeoning industry for the control of this disease. *Photo: Phil Lester*

survive for long periods they still have to be introduced into a hive in sufficient numbers to initiate an infection. Fifth, because beekeepers have been burning and destroying infected hives for a long time, effectively they have been trying to eradicate the disease already. Our laws require that all beekeepers register the location of their hives and apiary sites, and manage AFB in their colonies. Finally, feral colonies were considered not to have a large AFB problem. These feral colonies have also been devastated over the last two decades by *Varroa*.[47]

A goal of elimination and eradication is tantalising, especially with such a destructive disease. 'All we need is for 9300 beekeepers to take the following actions across their 924,000 hives located in 59,000 apiaries,' says Clifton King, who is the National Compliance Manager of the AFB Management Agency in New Zealand. 'We need to increase the frequency and efficacy of inspections, improve traceability [of honey, hives and hive equipment], improve destruction and sterilisation, and improve management practices to reduce the potential for AFB to spread from colony to colony.'[48] Sounds easy, doesn't it?

There are, however, several major challenges or problems that will hinder AFB elimination in New Zealand. Of all the diseases to try to eradicate, one of the most challenging pathogens must be one whose endospores can hide away for decades but still be viable. There is also the fact that this disease may be 'evident only after several years of "subclinical" disease' (and yes, that is the third time I've quoted Genersch's work).

Yet another obstacle for AFB elimination is apparent when we look at compliance by beekeepers. Beekeepers are a diverse group. A large proportion of beekeepers in New Zealand are willing and able to eliminate AFB from their hives. They are compliant and law-abiding. But a small proportion are anti-establishment, unable, and/or frankly don't give a damn. Many of these beekeepers will not make any effort towards AFB control and are recidivist offenders. I suspect this group has a variety of motivations. Some probably believe that AFB control is a lost cause, while others might have more selfish reasons. Some might be willing to eliminate AFB from their business but lack the required skills or knowledge, though this group can be helped by education and training.

The levy imposed by the government on beekeepers for AFB elimination is sufficient for the Management Agency to inspect 2% of apiaries in New Zealand each year. Their current practice is to focus on these recidivist beekeepers who know they have AFB but choose not to manage it. Potential offenders are indicated when clusters of other, law-abiding beekeepers report AFB. This clustering probably indicates a source population and allows AFB inspectors to focus their efforts on beekeepers in specific regions. The agency currently focuses on the causes of these clusters and identify recidivist beekeepers who have operations with >10% AFB infection rate in their hives.

One example of a commercial beekeeper who was knowingly operating a business with AFB demonstrates the problems that the Management Agency face. This beekeeper had approximately 900 beehives. Upon inspection, 32% of these hives displayed clinical and clearly identifiable symptoms of AFB. Nearly 90% of the beekeeper's apiaries were infected. Entire apiaries had been robbed out, causing honey bees from around the region, including those of neighbouring beekeepers, to become inoculated with this disease. The beekeeper had been recycling frames from diseased hives and thus spreading AFB between hives. His business had been harvesting honey from infected hives. It is easy for us to point the finger at this sort of beekeeper behaviour and urge authorities to throw the book at their offending. And there is justification for book-throwing, given the cost of AFB to neighbouring beekeepers and the ultimate goal of AFB elimination. On the other hand, many beekeepers in New Zealand are struggling financially. For this beekeeper, keeping his business running, albeit by poor management, was probably at the top of his mind. These situations are often complex, with multiple layers and sources of stress that lead to poor decisions and bad behaviour.

Worse still is the intentional AFB infection of bees. I've heard of beekeepers intentionally spreading AFB to competitors, though these reports are difficult to verify. There is a name for this practice: 'seeding' or 'spiking'. I've heard reports of beekeepers placing infected frames of honey and brood near their neighbour's apiaries, or throwing infected frames to near a competitor's apiary from car windows. Other tell me that some seeding occurs by dissolving diseased bees and honey into liquid and spraying this inoculum directly into hives. This behaviour appears more common now that immense amounts of money are to be made from mānuka honey. I have no idea of the legitimacy of these reports – or whether these actions resulted in the transmission of AFB – as nobody would willingly own up to seeding or spiking. But if these reports are true, then it seems that AFB is being weaponised.

One possible source of hope for the elimination of AFB has been the arrival of another devastating parasite and pathogen combination. Scientists have hypothesised that the arrival of *Varroa* into New Zealand and other countries will help with AFB management. A key 'benefit' of *Varroa* is that feral honey bee hives are substantially reduced or even eliminated. While it is thought that feral hives are rarely infected with AFB, the near absence of feral hives should mean that unmanaged sources of this pathogen are very rare or are even non-existent.

I'm not quite so optimistic regarding the benefits of *Varroa* for AFB control. One Kiwi beekeeper I know tells me that he is aware of a feral nest site high in a tree in a forest near one of his apiaries. It's a favourite site for swarms that have escaped his clutches. Like all beekeepers he attempts to limit or capture any swarms, but sometimes one gets away – or, in his view, 'more probably' a neighbouring (and possibly less competent) beekeeper loses a swarm and the swarm then finds that nest site in the tree. The beekeeper tells me that AFB is sure, then, to turn up in his apiary.

That scenario is just one beekeeper's experience and perception. But a lot of it resonates with what we know about the dynamics of AFB. Yes, the vast majority of feral, unmanaged colonies will likely eventually die because of *Varroa* infestations. A colony's death, however, won't be instantaneous. An escapee swarm will occupy a feral nest site for many months. Swarms are usually produced in spring, and the feral hive then grows until late summer. A typical *Varroa* infestation will then build in number and weaken the colony, perhaps killing it in autumn or the following winter. Those months between a spring colonisation and colony death provide ample time for an AFB infection to develop, especially if that hive site is preloaded with endospores from the previous occupants. You'll remember that AFB endospores live for decades. So even if it is five years since that feral nest site was occupied, there is still substantial infection risk. That *Varroa*-weakened hive, with its honey reserve, would then be susceptible to robbing and spreading of AFB in autumn, especially if the feral hive is nearby managed hives. As the authors of the Finnish study concluded, robbing is undoubtedly 'one of the most common and serious transmission routes of *P. larvae* spores [AFB endospores] within apiculture'.[49]

A big caveat is that I don't know for certain if this feral colony–*Varroa*–AFB interaction actually occurs, and, if it does, how often. Historic data suggests that 6.4% of feral colonies within New Zealand were infected with AFB.[50] AFB infection from a feral colony site might be a rare event. Nevertheless, with a management goal of elimination, even the rare event of AFB infection in this way is a major spanner in the works. I'm sure that *Varroa* has reduced the number of feral hives in New Zealand and around the world, which seems likely to help with AFB control. But my suspicion is that *Varroa* doesn't provide quite the benefit to AFB control that many might hope.

I think eradicating AFB in New Zealand by 2030 will involve some big challenges and at a minimum will be extremely difficult. The substantial challenges for AFB elimination, as I see them, include (i) endospores that can survive decades or even millions of years but still be infectious, which means that old feral hive sites or disused equipment are a concern; (ii) AFB typically goes unnoticed in a hive for years prior to its displaying 'clinical' symptoms, especially in ERIC II strain infections; (iii) some hives can even host this pathogen while appearing to be asymptomatic; (iv) the massive increase in the number of hives and apiaries in New Zealand is bad news, as classic epidemiology predicts that disease presence and prevalence increases with increasing host abundance; and (v) clearly not all beekeepers, commercial and amateur, are engaged with the goal of AFB elimination, and some appear to be actively spreading it.

AFB elimination is going to be tough. The biology of AFB means it will be hard to eradicate this disease at the level of an entire country. New Zealand is not currently engaging with Clifton King's goal, whereby all 9300 beekeepers would take action across the 924,000 hives located in 59,000 apiaries in New Zealand. If AFB couldn't be eliminated from managed hives on the 118 square-kilometre island of Jersey, it will be especially challenging on the 268,021 square kilometres of New Zealand.

I liken New Zealand's AFB elimination plan to the Predator Free 2050 campaign. The goal of this campaign is to rid New Zealand of the most damaging introduced mammalian predators – rats, stoats and possums. Nearly everyone would love to see this achieved. It would have major conservation benefits to our island nation. The benefits to our endemic flora and fauna would be huge, and we would see economic returns from tuberculosis management (possums host and spread tuberculosis). Similarly, the elimination of AFB would benefit the beekeeping industry and the country as a whole.

The complete eradication of pests, however, is extremely costly and requires massive investment. At a recent conference, Dr Bruce Warburton, of the government-owned research institute Manaaki Whenua Landcare Research, said that eradicating pests by 2050 was 'just a dream'. He highlighted differences between control and elimination or eradication: 'There's a huge difference between a 98 per cent kill and a 100 per cent kill . . . A 95 to 98 per cent kill with 1080 [the key toxin or tool used for pest control] operations cost NZ$30–$40 a hectare, but a 100 per cent kill and total eradication would cost NZ$200,000 to $500,000 a hectare.'[51] The difference between control and eradication is huge, at least with current techniques and approaches for pest management. With the development of new tools and technologies, that price per hectare may become more affordable and achievable. Accordingly, the New Zealand government is investing in the development of novel pest control tools.

As with Predator Free 2050, 'control' seems much more likely than 'elimination' for AFB. If elimination is to occur, I think much more investment is needed. Perhaps not the 10,000-fold-increased cash injection that Bruce Warburton has described, but I'm guessing a substantial funding boost to the order of some magnitudes would be required. We need more people on the ground searching for AFB. We need better compliance from beekeepers, perhaps with harsher penalties for having or spreading this disease. The beekeeping industry needs to support this goal. Better detection tools and epidemiological models are an absolute necessity. All cases of AFB must be given high priority – for everyone, not just the recidivist large commercial operations. Given that the target date of 2030 is less than 10 years away, New Zealand needs these resources and tools very quickly. And, even with all that, because endospores can persist and survive millions of years, the elimination of AFB might also be just a dream.

I'm sure that the proponents of New Zealand's AFB National Pest Management Plan, whose goal is to eliminate American foulbrood in managed colonies, will point to those large commercial beekeepers who apparently have eliminated AFB from their operations. Perhaps elimination really is achievable – I hope so.

5. PATHOGENS

Fungal, trypanosomatid and other parasites and pathogens

In the spring of 2014, beekeepers on the Coromandel Peninsula of New Zealand reported a mass mortality of bees. Thousands of hives were lost, resulting in a 40–65% reduction in honey production for many commercial companies in the region. Reports flooded in of beekeepers observing the loss of nearly all bees in many of their apiaries and hives. Their bees had disappeared without a trace. Beekeepers from nearby and distant regions were also losing colonies, with the same symptoms and devastating effects for their businesses.

The mass-killing symptoms involving the sudden loss of around 95% of bees in the colonies, leaving just the queen and a small number of worker bees, made this epidemic sound like the dreaded Colony Collapse Disorder (CCD). Mark Goodwin was one of the Kiwi scientists attempting to isolate the cause. He recognised that the epidemic was 'in some ways very similar to colony collapse disorder. [But] the difference is in the timing'. CCD usually occurs in autumn, whereas the Coromandel bee losses occurred in spring. 'Whether that is significant or not, we don't know,' he said. Were pesticides to blame? Coromandel commercial beekeeper and scientist Oksana Borowik had that exact thought. Evidence of toxic pesticides, however, was not to be found, and the widespread nature of the bee mortality also suggested that pesticides were unlikely to be responsible. New Zealand agriculture provides a relatively diverse landscape of crops and cropping practices. It seems unlikely that bees in so many different apiaries and regions would be suddenly exposed to the same pesticide.

Another hypothesis was that pathogens had caused the deaths. In support of this theory, high levels of three pathogens were identified from the surviving bees and hives. These bees had substantial infections of a trypanosome parasite called *Lotmaria passim*. The surviving colonies on the Coromandel Peninsula also had very high levels of the fungal pathogens *Nosema apis* and *Nosema ceranae*. All three are parasites or pathogens that live in the gut of honey bees. 'We don't know if *Lotmaria* is killing bees but we have identified it as another parasite,' said Borowik. 'We're trying to identify what killed them and have to do further research to figure out what is going on, either in synergy with *Nosema ceranae* or on its own.'[1]

At the time of writing, five years have passed since the 2014 mass-mortality event. We still don't know for certain why those bees died and why so many hives collapsed over such a large area. Oksana still favours the pathogens hypothesis,

Beekeepers on the Coromandel Peninsula lost up to 95% of their colonies in the spring of 2014. This is one of the apiary sites that has since recovered, with beekeeper and scientist Oksana Borowik with daughter Kate Davies, keeping a careful eye on their bees. *Photo: Phil Lester*

with *Nosema ceranae* as the most likely culprit. It seems likely that this pathogen was, at that time, a recent arrival to New Zealand. The year preceding the outbreak was also especially wet and cold. The bees were already stressed in a region with high hive densities. 'We had individual bees with 100 million *Nosema* spores,' Oksana told me. 'We'd never seen anything like it before. That year and losing all those hives has changed the way we manage our bees: we now overwinter hives differently, with an increased emphasis on hygiene, and smaller apiary sizes.' Fortunately, such a wide-scale and high-mortality event hasn't been seen in New Zealand since that time.

In this chapter I'll talk about some key trypanosome and fungal pathogens or parasites. I'll introduce these species and talk about how they have influenced bee populations. Trypanosome parasites in particular have been linked with dramatically declining wild bee populations around the globe. *Nosema ceranae* has been linked with CCD in North America and countries including Spain. I'll finish by discussing how these and other pathogens and parasites can team up together to affect bee mortality.

Much of the evidence for the global, ongoing and devastating impact of trypanosome and *Nosema* pathogens comes from bumble bees. As we will see, these pathogens have been identified as one of 15 'emerging issues likely to affect global biological diversity, the environment, and conservation efforts in the future', as concluded in a 2017 'horizon scan' by an international team.[2] So for the next few pages we will drift away from honey bees, as I think that their large, furry relatives can deepen our understanding of the global effects and management of these pathogens.

The plight of the bumble bee: Trypanosome and fungal pathogens

Wikipedia (a friend of teachers and academics everywhere) will tell you that Trypanosomatida is 'a group of kinetoplastid excavates distinguished by having only a single flagellum'.[3] I'll try to translate. Their name is derived from the Greek *trypano* (borer) and *soma* (body) because of their corkscrew-like motion. They have a flagellum, or tail-like organ, that they wiggle around to propel themselves.

Entomologist Jay Evans (left) and Ryan Schwarz of the Agricultural Research Service in the US examine spores of *Nosema ceranae*. These scientists have shown how infections by trypanosome parasites and *Nosema ceranae* elicit distinct immune responses in honey bees. The immune system responds to parasite infections within six hours. *Photo: Science Photograph Library / Alamy*

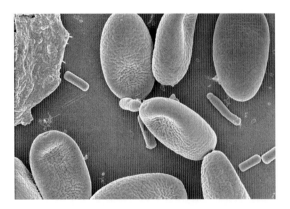

Spores of *Nosema ceranae* viewed through a scanning electron microscope. This pathogen is the jelly-bean shaped, rough-looking cells. They are about 3 μm (0.003 mm) long. There were up to 100 million of these spores per bee, in bees sampled by by Oksana Borowik during the collapse of her hives in 2014. *Photo: Peter Neumann*

Another thing that distinguishes this group is that their DNA is stored in an organ called a kinetoplast, which is very different to how your DNA is stored inside the nucleus of each cell. All trypanosome species are parasites and the majority are parasites of insects. The trypanosome parasites that you may have heard of are those in the *Leishmania* genus, which cause the disease Leishmaniasis in humans via blackflies or sandflies. Somewhere between 4 and 12 million people are currently infected with Leishmaniasis in 98 counties in Asia, Africa, South and Central America, and southern Europe. This disease, also known as white leprosy and black fever, causes up to 50,000 deaths per year. And just as with the endospores of the bacterial disease American foulbrood, relatives of this parasite have been found in fossilised sandflies from 20–30 million-year-old Dominican amber.[4] These parasites have been living off insects and a wide variety of animal hosts, including us, for a long, long time.

The trypanosome *Crithidia bombi* was discovered and first described in bumble bees in 1988. Over the last 30 years we've learned a lot about this parasite. After its cells have been inadvertently eaten or introduced into the bumble bee's digestive system, the parasite attaches itself to the gut wall of its host. Within the gut the parasite multiplies. Hundreds or thousands of new parasite cells are then excreted each time the bumble bee defecates. Bees aren't as tidy as you might think, and will, unfortunately, defecate in their nests and on resources such as flowers. While foraging for nectar or pollen from a flower, workers from an uninfected colony might come into contact with these infectious bee faeces and pick up parasite cells on their hairy legs, abdomen or mouthparts. Bees are always infected with *Crithidia* by this faecal–oral route. An entire colony can become infected from one infected individual. Daughter queens from these heavily infected nests are produced, are rapidly exposed to high levels of parasite cells, and quickly become preloaded with *Crithidia* infections for their new nests in the following year.

What happens to a colony that is infected by *Crithidia*? First, it's bad news for daughter bumble bee queens that carry an infection. They have to overwinter and attempt to start a colony in the following spring. Hibernating, overwintering queens that have a *Crithidia* infection lose more weight than their uninfected counterparts, their colonies are 40% more likely to fail and die in the spring and, if they do survive, their colonies are smaller. They produce fewer males and overall have fewer reproductive caste members, indicating a lower relative 'fitness'.[*5] The effects on individual forager bees are even more complex. The parasite instils learning deficits and generally slows the bee down, making them dull and sluggish.[6,7] Infected bees forage and move between flowers more slowly and take longer to learn about associations between flowers or flower colours and nectar rewards. Heavily infected bees can show a massive 200% delay in their learning capacity. It's been hypothesised that these stupefying effects are probably due to the high energy costs to the bumble bee as its immune system tries to fight the infection.[8]

But here is where it gets really interesting. The effects of the *Crithidia* parasite on bumble bee colonies aren't consistent. For a long time, people thought that *Crithidia* had little impact on bumble bee colonies. Similarly, the first description of a trypanosome pathogen from honey bees was in 1967, from apiaries in Australia. The species *Crithidia mellificae* was isolated and cultured, but was described as having 'no pathologic effects in the hymenopteran host'.[9] Since then, however, two factors have emerged as major influences on the effects of this parasite on bumble bee populations: stress, and strain-specific interactions.

Stress in bees often involves an inability to access a plentiful diet containing high-quality pollen and nectar sources, but also involves pesticides, other pathogens, competition and climate. First, let's deal with diet. The saying 'You are what you eat' is as true for bumble bees as it is for us. Dietary changes or limitations cause considerable stress on bees: a hungry bee is probably not a happy bee. If the bee has a restricted diet, *Crithidia* infections are much more virulent and damaging. In one experiment, groups of bumble bees that were infected or uninfected with *Crithidia* were fed either a normal diet or reduced rations (i.e. they were starved). The authors reported, 'Under favourable conditions the infection caused no mortality, while when hosts were starved the infection

* As an evolutionary or ecological term, 'fitness' refers to the number of offspring or reproductive success of individuals. Scientists often refer to relative fitness, which is the reproductive contribution of one group of individuals compared with others in the population. The influential biologist Maynard Smith referred to fitness as 'a property, not of an individual, but of a class of individuals'. In this chapter, we are interested in groups of individual bees infected by parasites versus a group of uninfected bees.

increased the host mortality rate by 50%'.[10] The amount of food available to bumble bees was a deciding factor in the effects of *Crithidia*.

Dietary limitations or changes also seem to have a range of other more subtle effects that are not obvious to the human eye. It's not just the amount of food they eat; the type and diversity is important as well. The species of pollen the bees eat has a substantial influence on *Crithidia* infections. A diet of pollen from the plant buckwheat can result in a 40-fold increase in the infection load of *Crithidia* parasites, compared with pollen from sunflowers.[11] This tells us that what bees eat is really important. It makes me want to run out and plant more sunflowers. Chemicals in the nectar of some plants can also be toxic to and inhibit the growth of *Crithidia*. Thyme, the common herb you might have in your garden, produces thymol, which can inhibit the growth and development of this parasite within bumble bees at concentrations as low as 4 parts per million in nectar.[12]

We are unsure why dietary stress makes *Crithidia* infections so much worse for bumble bees. The authors of the work showing how different pollen influences *Crithidia* infections suggested a range of mechanisms, including the rough sunflower pollen scouring the bumble bee gut or somehow preventing the parasite's attachment.[13] But the beneficial effects of pollen could also be indirect, as the bee's diet influences its natural gut bacterial diversity. Bumble bees have a distinct, resident bacterial community that can protect their furry hosts from *Crithidia* infection. Bees emerging from their pupal stage need to be exposed to faeces from their nest-mates in order to acquire these bacteria and develop a protective microbiota.[14] Effectively this is a 'faecal transplant' – like the transfer of stool from a healthy human donor into the gastrointestinal tract of a sick patient for the purpose of treating diseases such as ulcerative colitis. Bees have been doing this for millions of years, while humans have only just cottoned on. These mutualistic bacteria have been described as the 'extended immune phenotype' of their bee hosts, with groups of bacteria clearly being beneficial to bees by helping fight pathogens. Lower or reduced infections of *Crithidia* can be seen in bees that have a high microbiome diversity with large gut bacterial populations, including some of the bacterial groups known to be beneficial to people, such as *Lactobacillus*.[15]

Crithidia can clearly be bad news for bumble bees, particularly when they are stressed. It's true that bees aren't always stressed, especially in times of high pollen abundance and warm temperatures, but a bee colony that survives for several months will eventually suffer from some degree of strain. Pollen will become limited at some stage, and sunflowers don't grow everywhere all the time. Pesticides or herbicides might be sprayed onto a nearby crop, perhaps exposing the bees to sub-lethal doses. Stress will occur for most colonies at some time and this stress will affect the *Crithidia* infections that the bees carry.

From this description of *Crithidia* you'd be forgiven for thinking that bumble bees everywhere are about to be wiped out. Surely populations wouldn't persist if colonies had 100% of individuals infected with a parasite that can easily move between nests?

The saving grace for these bumble bees, it seems, is that different strains of the parasite have different effects on different genotypes or strains of bumble bees. It is thus likely that, as concluded in a 1998 study, these 'genotype–genotype interactions are the prevailing forces in structuring the interactions' between *Crithidia* and bumble bees.[16]

So, populations of bumble bees in one area that have high genetic diversity are likely to have some colonies that are resistant to a particular strain of this parasite. As long as you have these high levels of genetic diversity, population persistence is ensured. But strains of this parasite might be devastating to populations of bees with limited genetic variability, which is likely to be problematic for bumble bee species that have experienced massive declines. The great yellow bumble bee (*Bombus distinguendus*) was once distributed throughout the United Kingdom but now has much smaller populations in a fraction of its former range.[17] Parasites and pathogens are a major concern for the great yellow bumble bee and many others that have low genetic diversity and tiny distribution.

The picture that emerges from *Crithidia* in bumble bees is complex and hazy. It is not as simple as 'This parasite is bad for bumble bees'. Its effects are dependent on stress and on the genetic strain or genotype of both the bee and the pathogen. The pollen diet of the bee can mitigate or worsen the effects. Limited sources of nectar and pollen can turn a *Crithidia* infection from a puppy dog into a pit bull. Pesticides and immune reactions offer further complications. These complex interactions shape the health of bees.

A similar story emerges for a common fungal pathogen that infects bumble bees, *Nosema bombi*. This parasite was first described from these 'humble bees' in a 1914 publication.[18] Just as for the trypanosome parasite, *Nosema bombi* is transferred to bees by the faecal–oral route. It was also initially thought to be to be relatively benign as a pathogen, but recent experiments have shown it to be severely damaging. *Nosema bombi* infects not only the gut of bumble bees but also the fat bodies, which just as in honey bees are extremely important to immune function. *Nosema bombi* also infects the ovaries in the queens, and testicular organs in males.[19] The colonisation and parasitism of the reproductive organs is disastrous, as it can completely sterilise the bees. Workers are also affected and have reduced survival. Experimentally infected colonies placed in the field are smaller and generally fail to produce reproductive bees.[20]

The consequences of these bumble bee diseases have been extreme, especially in areas where these pathogens have been introduced along with exotic bumble bee species. In both South and North America we have seen dramatic declines in

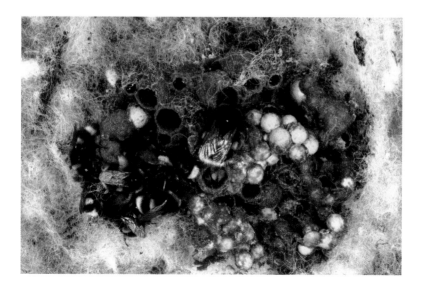

A nest of the buff-tailed bumble bee (*Bombus terrestris*). Compared with those of honey bees, these nests are small and look quite disorganised. There are only small pots of honey. The cells contain larvae and pollen. The queen is shown to the right of a cohort of worker bees. *Photo: Phil Lester*

native bumble bee abundance and diversity, apparently due to the introduction of the common, buff-tailed bumble bee, also known as the large earth bumble bee (*Bombus terrestris*). Each year over a million colonies of this bee are commercially produced, bred and introduced around the world for pollination purposes.[21] *Bombus terrestris* has been used in Canada, Japan, Korea, North Africa, Russia and the United States for the pollination of greenhouse crops such as tomatoes. Here in New Zealand it has been commercially produced for pollination since the 1990s. New Zealand has exported these bumble bees to China and currently even to the Maldives, a small coral-atoll nation to the south of India. For many years the buff-tailed bumble bee was produced in Europe and exported into North American glasshouses for pollination. Though these introduced bees are taken into commercial glasshouses, escapes can occur. Growers might want the bees to focus just on their plants, but bees won't be told where they can't forage. The introduced colonies find outlets and interact with the native pollinator community. One study in Canada found 73% of the pollen on bumble bees returning to hives inside greenhouses was from flowers found only outside the facility.[22]

When bumble bees are exported, so are their pathogens and diseases. *Nosema bombi* appears to have been just such an export, with massive negative effects for recipient communities of bumble bees. A widespread pattern of native bumble bee decline was observed after the buff-tailed bumble bee was imported into North America from Europe. These native bumble bees have declined by not just a little: they declined in abundance by up to 96% and contracted in range by 23–87% over a 20-year period.[23] *Nosema bombi*, introduced from Europe with buff-tailed bumble bees for pollination, has been indicted as a key player in this decline. The introduced bees escape from their greenhouses and spread pathogens and parasites to the native pollinators. These native species are often immunologically and evolutionarily naïve to these parasites, hence their devastating effects.

This pattern has been repeated in South America. The buff-tailed bumble bee and the large garden bumble bee or ruderal bumble bee (*Bombus ruderatus*) were brought to Chile for pollination purposes. The populations of ruderal bumble bees were introduced from New Zealand, where they had been colonised after shipment from England, because they were assumed to be relatively free of disease. The buff-tailed bumble bees likely came from Israel and Europe. Subsequently, wherever the buff-tailed bumble bees appeared, it became the dominant bee and native bumble bees rapidly declined. Another of the most extreme of these effects has been via the severe decline and local extinctions of the sole native Patagonian bumble bee, *Bombus dahlbomii*.[24] The introduced bees competitively displace this Patagonian bee, damage flowers by their foraging, and transmit pathogens as well. For these South America pollinators it seems that *Crithidia* plays the major role in decline and replacement, but *Nosema bombi* is there as well.[25]

So how much of a problem is the introduction and movement of bumble bees and their diseases around the world? As mentioned earlier, a 2017 analysis identified invasive bumble bees and the diseases they vector as one of 15 'emerging issues likely to affect global biological diversity, the environment,

The Patagonian bumble bee (*Bombus dahlbomii*) on lavender flowers. This South American pollinator has had its range substantially reduced and is now threatened with extinction after the introduction of bumble bees such as the buff-tailed bumble bee. Introduced bees competitively displace the Patagonian bumble bee, damage flowers by their foraging, and transmit pathogens. *Photo: Michael Grant Wildlife / Alamy*

and conservation efforts in the future'.[26] Much like climate change and the use of plastics, a globally coordinated response is needed. A harmonised approach is especially important for neighbouring countries. For example, Chile still to this day imports and spreads buff-tailed bumble bees, while their neighbour Argentina maintains a policy of rejecting requests to import alien bumble bees for commercial use, based on risk assessments. Argentina has instead developed one of their native species for commercial rearing and pollination purposes. Bee researchers in this region are immensely frustrated over the wildly differing approaches for managing and conserving their bees.[27]

The plight of bumble bees is a globally important conservation issue. It also highlights just how much of a problem *Nosema* and trypanosome pathogens can be for bee communities.

Nosemosis, caused by *Nosema*, in honey bees

The famous French microbiologist Louis Pasteur had an amazing career that touches the world on a daily basis. His contributions include inventing the process of pasteurisation, discrediting theories about the spontaneous generation of microorganisms, carrying out some immensely unethical experiments with vaccines that have gone on to to save many millions of lives, and inventing processes for controlling beer fermentation. Pasteur's legacy also involves *Nosema* pathogens in insects. In 1870 he produced a book on the 'pebrine disease' of silkworm, which is caused by *Nosema bombycis* and appears as brown spots or corpuscles over the body of these caterpillars.[28] He proposed ways to control this disease, but despite his efforts the French silkworm and silk industry never recovered from these disease epidemics.

These tiny, single-celled, bean-shaped organisms live inside their host's cells. As you can see in the scanning electron micrograph of *Nosema*, they have no legs or tail. They cannot move. The spores, or infectious stages, are reliant on their hosts to inadvertently eat the parasite. Once in the gut, germination is initiated by specific conditions such as the pH and ion composition. Filaments extend from the parasite, which then forcibly penetrate and inject the spore contents

The Isle of Wight disease, showing the queen among a small cluster of dead bees on comb, taken from the center of the brood-nest in early spring. *Photo: Library Book Collection / Alamy*

into the gut cells of their hosts. Once in the cells the parasite absorbs the sugars present, multiplies, and spreads to other cells within the midgut. These energy-eating parasites then reproduce within the cells and produce a type of spore suited to infect other cells within the bee gut. More than 200 million of these little parasites can be found within a single adult honey bee. At some point, tough spores are produced that burst their host cells and exit the insect with its faeces. The cycle starts again after the spore is inadvertently eaten or introduced into the digestive system by a new bee host.[29]

Nosema are one of the most widespread and prevalent of all the pathogens and diseases found in honey bees.[30] Of the two species infecting honey bees, *Nosema apis* and *Nosema ceranae*, we've known about *Nosema apis* for the longest. This pathogen and the Nosemosis it causes in honey bees was first described in the early 1900s. The infamous Isle of Wight disease was initially attributed to *Nosema apis*. An interesting array of solutions were suggested to control *Nosema* on this island, as reported by *Nature* in 1917: 'A preparation of coal tar, a combination of several germicides, hydrogen peroxide, sulphate of quinine, and even pea-flour have all been put forward as sovereign remedies and extensively sold to distracted beekeepers.'[31] Bizarre remedies and distracted beekeepers aside, *Nosema apis* has been with the beekeeping industry for thousands of years.

It hasn't been – and probably won't be – a major source of mortality for honey bees. It probably wasn't the primary cause of the Isle of Wight disease, which instead was likely to be due to a variety of factors including the overstocking of bees, inclement weather that inhibited foraging, mite parasites, and viruses. However, it would be impossible to exclude *Nosema apis* as a contributing factor to poor bee health on the Isle of Wight in the early 1900s, or in the current day.

The official name for the disease caused by *Nosema* in honey bees is Nosemosis. A typical seasonal trend is for a low abundance of the disease in summer, followed by rising levels in autumn and a peak in winter and spring. To the beekeeper, one of the most obvious symptoms of a *Nosema apis* infection is that the bees have diarrhoea. Beekeepers sometime refer to it as 'bee dysentery'. The photograph below is of just such a beehive, with bee diarrhoea on the snow and outside hive walls as the bees emerge for the first time in spring. Gastrointestinal upset is typically most obvious after a period of cold weather, or in spring, when the bees emerge en masse and defecate runny, yellow-brown stripes on the outside walls of the hive. Worker bees seem most affected, possibly because they are most frequently exposed to *Nosema* spores as they clean cells in the hive. Workers are weakened and their longevity is reduced. The colony's honey production declines. Drones and queens seem less commonly afflicted by the disease, but if the queen does become infected then her ovaries will degenerate, resulting in reduced egg-laying. Workers will sense

Historic photograph showing dysentery on the hive walls and on the surrounding snow in late winter, likely indicating a *Nosema apis* infection.
Photo: Library Book Collection / Alamy

her reproductive slowdown and initiate queen supersedure.

Chronic infection in workers results in disjointed wings and an inability to fly. *Nosema apis* also impairs the bees' ability to digest pollen. They have swollen abdomens and their stinging reflex is altered. Other effects include the atrophy of the hypopharyngeal glands – the long, coiled glands on the sides of the bee's head that are important for the production of royal jelly.

Nosema apis has been parasitising honey bees for millennia. *Nosema ceranae*, by contrast, appears to be a much more recent and emerging threat to honey bee health and wellbeing. This pathogen has been correlated or associated with Colony Collapse Disorder. The effects of *Nosema ceranae* on bees and bee hives are quite different from those of *Nosema apis*. The only obvious symptoms are that the bees appear weakened, are sometimes unable to fly, and their honey production is lowered. Unfortunately, these are also symptoms of many honey bee diseases. With *Nosema ceranae*, the telltale bee dysentery of *Nosema apis* is absent.

All the honey bee colony members are susceptible to *Nosema ceranae*. The drones appear especially susceptible and have even been named intracolonial 'super spreaders', because pathogen transmission seems to be enhanced when drones rather than workers are the infected individuals within a hive.[32] It's bad news to keep infected drones near workers. Much less obvious to the beekeeper is that infected bees' immune systems are suppressed, increasing the bees' susceptibility to other pathogens. *Nosema ceranae* seems to cause bees to have trouble synthesising protein, as well as using and synthesising fat. The bees also use and seek more carbohydrates. It has been hypothesised that these changes are induced in part because the parasite destroys the bees' gut lining and the bees are stressed. Their foraging and metabolic rates change accordingly, which results in the observation that infected bees live relatively shorter lives.[33]

The species name *ceranae* tells us about the origin of this parasite. This pathogen was first described from the eastern honey bee, also known as the Asiatic honey bee (*Apis cerana*) in 1996.[34] The bee samples were from the Bee Institute of the Chinese Academy of Agricultural Sciences outside Beijing. *Nosema ceranae* has since been found to be able to infect European or western honey bees (*Apis mellifera*) as well as the dwarf Asian honey bee (*Apis florea*), the giant Asian honey bee (*Apis dorsata*), and Koschevnikov's honey bee (*Apis koschevnikovi*), which is found in Malaysian and Indonesia. There are even reports of *Nosema ceranae* affecting three native South American bumble bee species, which are now all in decline.[35]

Once *Nosema ceranae* was described and defined, people started looking for it all around the globe. The earliest known bee samples found to be infected by it date back to 1979, from Brazil.[36] In the United States it was also found in historic samples from the mid-1990s, and in Uruguay in one sample from pre-1990.[37] A variety of strains are known to be distributed around the world, including in New Zealand.[38] We don't know when European honey bees became infected, but clearly it has started to parasitise these new hosts over the last several decades. Just as with *Varroa*, humans shifting bees around the globe and into the range of the Asiatic honey bee have almost certainly led to the parasitism and global infection of *Nosema ceranae* in European honey bees.

Nosema ceranae's effects on western honey bees are probably so considerable because the pathogen didn't co-evolve with this species. Parasites and pathogens that have lived with their hosts for thousands of years appear to be relatively benign: if you depend on your hosts (and future generations of hosts) to live, it pays not to kill them. Consequently, in any population, over time the nastier strains tend to die out. This leaves the less damaging individuals to become the strains that successfully reproduce and infect future generations. It is often when a pathogen jumps from one species to another that we see the devastating effects.

This host naïvete is probably why *Nosema bombi*, which appears to have co-evolved with the buff-tailed bumble bee, has such a harsh effect on North and South American bumble bees. The new host species are immunologically naïve. Similarly, a species of *Nosema* that is found in the invasive harlequin ladybird beetle has little or no observable effects on its co-evolved host. The beetle appears to have an unusually strong antimicrobial immune system. But if the beetle invades a new country or continent, the parasite infects new, native ladybird beetles, quickly killing off new host species that would otherwise attack or compete with the invasive predator. This *Nosema* has been described as a biological weapon that selectively kills native ladybird beetles and facilitates the invasion and success of the harlequin ladybird beetle.[39]

At the level of the honey bee colony or hive, the effects seem to vary according to geography.[40] Early studies in Spain showed a highly virulent parasite that ultimately would lead to colony collapse. Similarly, honey bee colony collapse was seen in Greece and Israel after the *Nosema ceranae* infection was found. Studies in North America found no correlation between colony collapse and the presence of *Nosema ceranae*, but they did show that this parasite was 'over-represented', or more common than you'd expect by chance, in collapsing colonies. In countries such as Uruguay or Germany there is no correlation between this pathogen and colony mortality. Scientists think that countries with colder climates 'do not fulfil the specific conditions (climatic and/or beekeeping practices) for *N. ceranae* to compromise colony survival'.[41]

It certainly seems that climate interacts with pathogens such as *Nosema ceranae*. Different genetic strains and levels of bee genetic diversity also seem to change the outcome of infection.[42] That the effects of this parasite on hives and colonies vary from country to country is, I think, at least partly due to the immunological suppression of *Nosema*, in combination with differing strains of other diseases. The bumble bee research I discussed earlier tells us there are complicated interactions among the genetic diversity of the bee host, the strain

A harnessed worker is given sugar water with an infection of *N. ceranae* for experimental purposes. Despite being restrained, the bee displays the proboscis extension reflex – sticking out her tongue when her antennae are stimulated. She then drinks the solution. *Photo: Phil Lester*

of parasite, the climate, and probably other parasites and pathogens in the area. We know that different strains of pathogens such as *Crithidia* or DWV occur in different countries. Perhaps Spanish bees are exposed to a strain of DWV that becomes more lethal in combination with the immunosuppressant effects of the European *Nosema ceranae* populations. It is also possible that this *Nosema* species is made more pathogenic by its limited diet or its exposure to pesticides, or that it will become more virulent with climate change.[43] Currently, these ideas are hypotheses. It's very hard to demonstrate cause and effect here, as much of what we know comes from correlational studies. Correlations are notoriously unreliable: a simple correlation does not imply causation.*

Louis Pasteur controlled *Nosema* infections in his French silkworms by culling infected individuals and breeding only from females that showed no signs of infection.[44] Selective breeding might be a future possibility for *Nosema* control in honey bees and, as a 2012 study in Denmark has shown, it can be undertaken successfully in some areas,[45] though the priority for any honey bee selective breeding for the foreseeable future will be for *Varroa* resistance. At the moment we are very limited in our arsenal of management options. In some countries the antibiotic chemical Fumagillin, a chemical first extracted from the fungal pathogen *Aspergillus fumigatus* in 1949, is used specifically against *Nosema* pathogens. *Aspergillus fumigatus* periodically kills people, and the extracted Fumagillin antibiotic can be toxic too. Side effects of Fumagillin include a significant increase in the risk of cancer and chromosomal mutation – small issues that most of us like to avoid – and are reasons why many countries refuse to allow it anywhere near honey bees. It can also detrimentally alter the honey bees' gut bacterial community and reduce bee longevity.[46]

---

* My favourite example of a misleading correlation is the proven, strong, positive correlation between ice cream consumption and rates of human drowning. When a nation is eating a lot of ice cream, there are a lot of drownings. There are fewer drownings when shops aren't selling ice creams. Ban ice cream and save lives! Right? Wrong. The two aren't directly related. When it's hot, people tend to both eat more ice cream and expose themselves to the risk of drowning, by swimming. It could be similar with *Nosema ceranae*. Perhaps other diseases or parasites cause poor health in bees and over time cause colonies to collapse. But, while they're still hanging on, this immunologically weakened colony becomes susceptible to a range of pathogens, including *Nosema ceranae*.

Nevertheless, Fumagillin appears to be the most effective of the currently available 'chemotherapies' available for *Nosema ceranae* control.[47] But different approaches and concentrations appear to be needed for this pathogen species. While Fumagillin has been used against *Nosema apis* in the American apicultural industry for more than 50 years, there is evidence that the current application protocol in the United States for Fumagillin may exacerbate *Nosema ceranae* infections rather than suppress them.[48]

In addition to Fumagillin, though, there are emerging approaches or techniques that could be useful. The oxalic acid and thymol that many beekeepers use to control *Varroa* mites may help control *Nosema* infections.[49] Other suggested or emerging approaches include the use of aspirin, anti-ulcer drugs, some natural extracts, amino acids and proteins that appear to stimulate bees' immune systems, and gene silencing. Gene silencing appears to have great potential. It involves feeding bees short strands of RNA that are designed to inhibit the genes that are vital for *Nosema* to invade bee cells and then proliferate. A 2018 study demonstrated that the use of gene silencing could dramatically reduce *Nosema* infections, resulting in an increased bee lifespan, better health, and stronger immune gene expression. 'These results strongly suggest that RNAi-based therapeutics hold real promise for the effective treatment of honey bee diseases in the future, and warrant further investigation,' the authors write.[50] I think so too.

Crithidia and *Lotmaria* in honey bees

We've known since 1912 that trypanosome parasites attack and infect bees. The first detailed work on these parasites was published by Ruth Lotmar, a German-born bee researcher and zoologist, in 1946. She initially worked with Karl von Frisch, but because of her Jewish ancestry she was hounded out of Germany prior to the Second World War, fleeing to Switzerland to live and eventually work.[51] In her research she described diseased adult bees which displayed a unique gut scarring pathology. Ruth termed this disease 'Schorfbienen', which translates to 'scab bees'. These scabs were melanised scars specific to a region

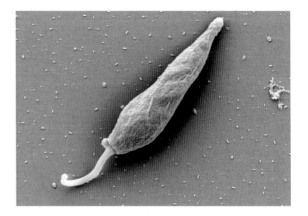

A scanning electron microscope image of the trypanosome parasite *Lotmaria passim*. These cells are about 4μm (0.004mm) long, excluding the flagella.
Photo: Peter Neumann

of the honey bee gut between the midgut and small intestine. Ruth's drawings and photographs clearly identify these pathogens as trypanosomatids. She didn't name the parasite in her publications, but her work was acknowledged and accoladed when a new genus and species of this parasite was named *Lotmaria passim* in her honour in 2015.[52] Two decades after her 1946 work, *Crithidia mellificae* was described from the honey bees in apiaries of Victoria, Australia.[53]

More recent research has demonstrated that both *Crithidia* and *Lotmaria* primarily colonise the hindgut of the honey bee. They are usually seen in the rectum. There, they attach to the gut wall via their flagellum or tail-like appendage. A little parasite colony forms, resulting in damage to the gut wall and the formation of the scabs that Ruth Lotmar described. Bees are infected by the faecal–oral route as worker bees clean the hive or forage on infected plant flowers. The damage done to the gut by *Crithidia* can substantially reduce the lifespan of honey bees and even solitary bees that are susceptible to this pathogen.[54]

Crithidia mellificae and *Lotmaria passim* are both globally widespread and abundant trypanosomes that parasitise honey bees. Again, their effects seem to vary around the world. Perhaps the most extreme case was found in a 2013 study of Belgium beehives led by Jorgen Ravoet. Of the 363 hives that were sampled, 46% died over winter, with *Varroa* mites and *Crithidia* contributing most to explaining winter mortality. The authors found 'a large effect of the occurrence of *Crithidia mellificae* in summer on later winter losses, even enhanced through *Nosema ceranae* co-infection'. They conclude that the 'protozoan *Crithidia mellificae* has been ignored for a long time, but the current data highlight it as a new putative key player in honey bee colony declines'.[55]

This work clearly shows that trypanosomatid parasites such as *Crithidia* can be a major cause of mortality. But, as the study demonstrates, even more concerning is how damaging pathogens can be when multiple species co-infect honey bee hives.

All bee colonies everywhere are typically infected with multiple pathogens. Within a hive we might find *Varroa* and viruses including the deformed wing virus, Kashmir bee virus, *Nosema* and others. Jorgen's analysis showed that colonies with just three pathogens would have a 6% chance of dying, compared with a 52% chance of mortality if six pathogen species were found. The co-infection of pathogens clearly spells trouble for bees. It also makes it difficult to attribute a cause to colony collapse, especially for beekeepers who lack the analytical instruments found in scientific labs. I suspect that the vast majority of beekeepers don't look for *Crithidia* and *Nosema ceranae*. A typical beekeeper will likely attribute colony loss to the most visible and obvious candidate, often *Varroa*. It seems to me that *Varroa* is still likely to be a major cause of hive mortality, but that pathogens like *Crithidia* are likely to contribute to overwintering colony collapse more often than we think.

Positive correlations between infections of *Crithidia* and overwintering colony mortality of honey bees have been observed elsewhere, including in North America.[56] North American bee colonies that exhibit Colony Collapse Disorder have been found to have 20-fold higher levels of *Crithidia* and a range of other pathogens, compared with healthy colonies.[57] But, just as in bumble bees, an infection of trypanosomatid parasites doesn't always mean death. That is why this pathogen was initially described as having no pathologic effects on its hymenopteran host. Scientists speculate that the infection and effects of these parasites are likely influenced by an array of factors including climate, the genetic variation of the host and parasite, exposure levels, and the treatment regimes and environment. These factors sound very familiar – they're similar to those that scientists have posited when working on bumble bees.

We have only begun to understand the effects of *Crithidia mellificae* and

Lotmaria passim in honey bees – indeed, it was only 2015 that *Lotmaria passim* was described. The scientists who named it concluded that it was by far the most prevalent of the two species in honey bees. They provide good evidence that the parasite described by Jorgen Ravoet in Belgium and in Europe was *Lotmaria*, not *Crithidia*. It appears that *Lotmaria* is widespread – present in China, Japan, the United States, New Zealand and Switzerland – and uses a wide variety of insect hosts.[58] As yet, we know too little about these parasites to say whether they have different effects or what control methods we could or should implement. Similar to *Nosema*, there is good evidence that plant metabolites like thymol, from thyme, can inhibit the growth of trypanosome parasites.[59] As with *Nosema* pathogens, scientists are also testing gene silencing methods.[60]

Pathogen webs in collapsing honey bee colonies

I began this chapter with the story of colony losses on the Coromandel Peninsula in 2014. Our immediate response to this die-off was to seek the cause so that we could prevent the same thing from happening again. Was it a specific pesticide? Which specific pathogen caused the deaths? A similar reaction occurred after the Isle of Wight bee losses in the early 1900s and the Colony Collapse Disorder events in North America earlier this decade: If we can find that single cause or reason, then we can make a targeted plan of action.

I think that all too often, when we see a widespread loss of bee hives or a colony collapse event, we look for a single smoking gun. It would be ideal to be able to point to one cause for the death of the bees. And it is true that some species, such as *Crithidia bombi* in bumble bees, can have a massive effect alone, especially on novel hosts. The introduction of bumble bees and *Crithidia* into South America and other regions is a global threat to our biodiversity.

But I suspect that honey bee collapses or mass mortality events are often due to a complicated web of pathogens that interact with other stressors. These stressors include climate, pollen or food availability, the genetics of the bees and the parasite, and agricultural practices such as pesticide use. In a 2012 review – from which I've borrowed the title for this section – researchers explored recent losses in honey bee colonies and concluded that pathogen interactions and environmental stressors are important components of bee disease and mortality.[61] It is also true that several parasites, including *Lotmaria passim* and *Nosema ceranae*, were found in the collapsing hives of the Coromandel Peninsula. They were probably at least part of the cause, and may even have been driving it, though other parasites such as *Varroa* mites and the DWV they carry will have contributed to the mortality.

It's impossible to exclude all pathogens and parasites from your beehives. Unfortunately, many of those pathogens and parasites will interact, and can be more lethal to honey bees working together than alone. The sub-lethal effects of multi-species infections differ too. For example, we know that when *Nosema ceranae* and trypanosome parasites co-infect a honey bee, they elicit an immune response that is very different from the influence of each pathogen alone.[62]

In order to effectively manage bee health, we need to manage multiple pathogen species. And, periodically, when multiple ugly stars align – a high abundance of certain pathogens plus stressors that include diet, climate or pesticides – we will see widespread bee mortality. I suspect that this scenario was exactly what occurred in the spring of 2014. This web of factors are probably also why it's so difficult to determine the cause of Colony Collapse Disorder, despite the huge amounts of money and effort that have gone into attempts to isolate the cause and prevent its future occurrence.

6. PESTICIDES

How do neonicotinoids, Roundup, organic pesticides and other chemicals affect honey bees?

In the spring of 2019, honey bees across rural Russia left their hives after a cold winter, just as they have done for thousands of years. But this spring was different. Farmland in 24 of the 85 Russian regions was scattered with hundreds of thousands of dead bees. Andrei Malykhin, a beekeeper north of Kursk, found his bees dead or dying. 'They were having spasms and convulsions, and slow, painful deaths. You simply can't explain how it looks. And you can't do anything. I made frantic calls to find out if there's anything I can give them, any antidote. There was nothing.'[1] The widespread loss of bees raised concerns over the threat of severe economic losses to Russian agriculture, which would reverberate for many years. 'If we lose the bees, everything will be affected,' said Arnold Butov, the head of Russia's beekeepers' union. 'We have to protect them as if they are holy.'[2]

The consensus by Russian bee experts and beekeepers alike was that pesticides were to blame. But there were other opinions too. The Russian Economic Development Ministry said it was too early to conclude what was causing the widespread bee deaths. According to the ministry, the accusations that pesticides were solely to blame were 'one-sided'. One consultant who had worked with large-scale Russian farms for two decades was reported as saying that he didn't see how there could be a 'wholesale catastrophic misuse of pesticides' that year. 'Yes, you are going to get farmers misusing pesticides, but not across all of these regions at once.'[3]

This sort of debate is a global phenomenon. Protesters dressed as bees demand the ban of chemicals and wave signs that urge politicians and the government to act now to 'save the bees'. A class of insecticides called neonicotinoids have received a great amount of attention, emotion and protest.

A protester from Extinction Rebellion helps block a junction in London as part of an ongoing protest to demand action by the UK government in April 2019. *Photo: Guy Bell / Alamy*

Indeed, pesticides, and particularly insecticides, can and do kill bees.

Given that bees are insects, it would be surprising if insecticides weren't lethal to them. Different insects respond differently to different pesticides, and I suspect that honey bees are, in general, somewhat more susceptible to insecticides than many other insects. When you squirt a housefly with fly spray, frequently your fly will stare back at you through the 6000 lenses of its compound eyes as if nothing unusual has happened. It feels like the fly is silently laughing. By comparison, a honey bee receiving a fraction of that fly spray will quickly die on its back and buzz no more. Perhaps of more significance is the fact that tiny, tiny amounts of many pesticides, present in parts per million or possibly parts per billion, can affect bee health. Exposure to minuscule amounts of these pesticides might reduce a bee's lifespan, cause it learning difficulties, or make it more susceptible to parasites and pathogens.

There is no doubt that pesticides can harm bees by lethal and sub-lethal means. But how much of an issue are pesticides for bee health, especially in comparison with pathogens and parasites? What is the evidence for their widespread and substantial effects?

In this chapter my aim is to step back from preconceived ideas about pesticide use. To me, the key questions include: What pesticides are honey bees exposed to, and at what levels? How frequently are bees suffering substantial harm from the use of these chemicals? And have pesticides been associated with widespread bee mortality, such as with the Colony Collapse Disorder observed in North America and Europe at the turn of this century? What do the scientific studies tell us about the lethal and sub-lethal effects of these chemicals? Finally, what is the evidence that we should implore the government to halt the use of chemicals such as neonicotinoids? I'll discuss neonicotinoids as a special case study. Along the way, we'll visit a group of activists and scientists in countries such as Switzerland who are determined to have a complete ban on the use of synthetic chemicals in agriculture.

Defining pesticides

First, let's get back to basics and define 'pesticides' and 'insecticides'. Pesticides are substances that are used to manage, repel or kill a range of pests – including insects, weeds, rodents, slugs and snails, and microbial species such as fungal species. They are typically synthetic chemicals, but they may also be based on natural products or substances produced by micro-organisms. Keep in mind that the term 'pesticide' is often used for a whole host of chemicals used to kill pests. Insecticides are just one type or group of pesticides, used to target insects specifically. There are many other '-cides', including herbicides, fungicides and parasiticides. These too fall under the 'pesticide' umbrella.

Pests damage or take a major portion of our crops. Wheat and rice, for example, together contribute just under 40% of the world's total calorie intake. Each year an estimated 22% of wheat and 30% of rice is lost globally to plant pathogens or insects.[4] The key pathogens include fungal species called rusts and smut. Other plant pathogens of these crops include blights, which are caused by Phytophthora, a group of microorganisms that triggered the Irish Potato

Famine of the mid-1800s. The insect pests include herbivores that bore into the stem of the rice plant, or army worms that chew on the leaves, flowers and kernels of wheat plants.

A huge array of pesticides are used to control these and other pests. In New Zealand the government's register of veterinary medicines, agricultural chemicals and vertebrate toxic agents lists 3452 products. A single farm or orchard will use a fraction of these chemicals, though some are used often. A farmer growing wheat, for example, might use several sprays of fungicide each growing season to control powdery mildew, smuts or rust.

Pesticides in bee pollen and hives

With such a large number of chemicals being used in agriculture, you'd expect to see chemicals in beehives too. If a fungicide is used on apple trees in an orchard, you might assume that it is in the bees or on pollen that bees forage from these plants.* Researchers have undertaken several extensive surveys of the insecticides, herbicides and fungicides found on pollen harvested by bees, on beehive material such as wax, and in bees themselves. The list of pesticides found in hives is long and diverse – it can go on for pages. One of the largest studies in North America

---

* Bees typically don't do what we tell them to do. One of the most striking results from studies on pesticides found in pollen returning to beehives is how infrequently bees forage in the places we'd like them to forage. Orchardist and growers employ beekeepers to move hives into specific crops for pollination purposes. When the crop is something like almonds and there are no other flowers or pollen sources available, bees will dutifully forage on almond pollen and carry it back to the hive. Foraging patterns change, however, later in the growing season when more choices are available. For example, the authors of a 2013 North American study moved hives into seven different cropping systems. None of the pollen subsequently sampled from hives in blueberry, cranberry, pumpkin or watermelon fields was from these target crops. In a cucumber crop, only 1.1% of the pollen sampled was from cucumbers. Instead, the bees flew off to collect from nearby wildflowers. Such displays of rampant disobedience by bees has been seen in many studies. Their widespread foraging is important, as the authors of the study suggest. 'Beekeepers need to consider not only pesticide regimens of the fields in which they are placing their bees, but also spray programs near those fields that may contribute to pesticide drift onto weeds.'

found 121 different pesticides or breakdown products (metabolites) of pesticides in beehives.[5] Within pollen samples – which are important because pollen is a food source for the hive – 98 pesticides were observed. An average sample had residues from seven pesticides, although as many as 31 pesticides were observed in one sample. These results are typical of what we see in many other countries.

So, unfortunately, we can conclude that bees are exposed to a wide array of chemicals. They have weird names that are hard to pronounce, like chlorpyrifos, chlorothalonil and imidacloprid. Table 1 attempts to summarise the most common pesticides found associated with pollen collected by bees or in beehives. It's not an exhaustive list. However, many of the pesticides were observed in a small number or percentage of the samples. Because these pesticides were rarely seen, they were unlikely to have widespread effects. Here, instead, I'm interested in the frequently observed chemicals that might harm bees on a larger scale.

A beekeeper with his beehives in an oilseed rape field. Seeds of this crop are treated with neonicotinoids prior to planting, resulting in the pesticide being expressed in the plant. The pollen and nectar that bees collect can then become contaminated. *Photo: Buiten-Beeld / Alamy*

Several of the studies (listed below the table) are freely available online if you are eager to see the full list and detail of their methods. I've also provided estimates of the LD_{50}, which refers to the amount of pesticide that would be lethal to 50% of a sample of honey bees.

What does this table tell us about which pesticides might be affecting bees widely? There are two key attributes to consider. Firstly, the percentage of hives found to have the pesticide is important. If only 3% of hives in North America show imidacloprid (a neonicotinoid) contamination, I'm a lot less worried about imidacloprid's effects in North America than I am about its effects in France, where, in a field survey, 49% of hives were seen to have residues of this chemical. The French survey, carried out from 2002 to 2005, shows that there is good reason to be concerned about the number of hives being exposed to these chemicals there. Secondly, the actual amount of chemical contamination is important. The pesticide chlorpyrifos is estimated to have an LD_{50} of 100 parts per billion (ppb) for bees that come into direct contact with the pesticide. In Taiwan, 42% of hives show average residues of 128ppb. Bees there seem likely to be directly affected by the concentration of this pesticide in pollen. If I were in Taiwan, I'd worry about the use patterns of chlorpyrifos a lot more than if I were in Italy. Although 30% of Italian pollen samples were contaminated by this pesticide, the contamination was an order of magnitude lower, at 10ppb.

The different studies shown in this table are enlightening. Fungicide and herbicide chemicals are very commonly observed in bee pollen. Consequently, fungicides and herbicides are in the bee bread that nurse bees feed to larvae. These results make sense, given the prominence of plant pathogens in food production and the frequency with which chemicals are used to control them. Three of the most common chemicals observed were pesticides beekeepers use to control *Varroa* mites: amitraz, coumaphos and fluvalinate. Other than chemicals for the control of these parasitic mites, the most commonly observed insecticides were chlorpyrifos and carbaryl. And, as discussed above, the percentage of neonicotinoid-contaminated hives varied substantially between countries and surveys.

Table 1: Common pesticides in pollen collected by honey bees

Pesticide	Oral LD$_{50}$ *	Contact LD$_{50}$ *	US (i) **	US (ii) †	France §	Italy #	Taiwan
Carbamates							
Carbaryl	2310	11000	11% (117)	33% (58)	8% 219	0% (1)	11% (83)
Cyclodienes							
Endosulfan		78700	28% (11)	44% (2)	6% (81)	0% (0)	0% (0)
Formamidines							
Amitraz		750000	31% (148)	56% (172)	0% (0)	0% (0)	0% (0)
Neonicotinoids							
Thiacloprid			5% (24)	11% (1)	0% (0)	0% (0)	0% (0)
Acetamiprid	145300	81000	3% (59)	17% (59)	0% (0)	0% (0)	1% (53)
Imidacloprid	37	439	3% (39)	17% (3)	49% (1)	13% (2)	6% (22)
Thiamethoxam	50	240	0% (0)	0% (0)	0% (0)		0% (0)
Organophosphates							
Coumaphos		240000	75% (180)	33% (2)	2% (925)	0% (0)	0% (0)
Chlorpyrifos	2500	100	44% (53)	39% (3)	0% (0)	30% (10)	42% (128
Phosmet	3700	10600	0% (0)	28% (799)	0% (0)	0% (0)	1% (102)
Diazinon	2000	2200	0% (0)	17% (1)	0% (0)	0% (0)	3% (239)
Parathion methyl			0% (0)	0% (0)	5% (25)	0% (0)	0% (0)
Pyrethroids							
Fluvalinate		2000	88% (95)	89% (42)	6% (487)	0% (0)	48% (915
Fenpropathrin		500	18% (15)	0% (0)	0% (0)	0% (0)	0% (0)
Esfenvalerate		2240	12% (8)	39% (17)	0% (0)	0% (0)	0% (0)
Cyhalothrin	27	439	11% (3)	39% (15)	0% (0)	0% (0)	2% (83)
Bifenthrin		150	0% (0)	17% (7)	0% (0)	0% (0)	2% (118)
Fungicides							
Chlorothalonil		1812900	53% (3015)	94% (4491)	0% (0)	0% (0)	2% (85)

esticide	Oral LD$_{50}$ *	Contact LD$_{50}$ *	US (i) **	US (ii) †	France §	Italy #	Taiwan ¶
zoxystrobin	250000	2000000	15% (21)	56% (60)	0% (0)	3% (8)	0% (0)
aptan		1080000	13% (434)	50% (977)	0% (0)	0% (0)	3% (161)
enconazole			0% (0)	0% (0)	10% (28)	0% (0)	0% (0)
andipropamid	2000000	2000000	0% (0)	0% (0)	0% (0)	20% (9)	0% (0)
erbicides							
endimethalin		498000	46% (45)	28% (5)			0% (0)
trazine		970000	20% (14)	6% (0)			0% (0)
letrolachlor	1100000	1100000	15% (13)	0% (0)			0% (0)
utachlor			0% (0)	0% (0)			3% (39)

Table 1 is ordered by frequency of common pesticide within each family observed in pollen samples, from one of the first and most extensive studies on this topic (Mullen et al. 2010) plus all neonicotinoids observed and common pesticides found in other countries. LD$_{50}$ values are in ppb (parts per billion) and are the estimated lethal dose to kill 50% of a sample of honey bees. % values are the number of pollen samples positive for the chemical. Values in parentheses are the average concentration observed in ppb in that country.

* LD$_{50}$ values are from either Mullin et al. (2010; full reference below) or N. Ostiguy et al., 'Honey bee exposure to pesticides: A four-year nationwide study', *Insects* 10, no.1 (2019).

** C.A. Mullin et al., 'High levels of miticides and agrochemicals in North American apiaries: Implications for honey bee health', *PLoS ONE* 5 (2010), e9754, doi: 10.1371/journal.pone.0009754

† J.S. Pettis et al., 'Crop pollination exposes honey bees to pesticides which alters their susceptibility to the gut pathogen *Nosema ceranae*', *PLoS ONE* 8 (2013): e70182, doi: 10.1371/journal.pone.0070182

§ M.P. Chauzat et al., 'A survey of pesticide residues in pollen loads collected by honey bees in France', *J. Econ. Entomol.* 99 (2006): 253–62, doi: 10.1603/0022-0493-99.2.253

S. Tosi et al., 'A 3-year survey of Italian honey bee–collected pollen reveals widespread contamination by agricultural pesticide', *Sci. Total Environ.* 615 (2018): 208–18, doi: 10.1016/j.scitotenv.2017.09.226

¶ Y.S. Nai et al., 'Revealing pesticide residues under high pesticide stress in Taiwan's agricultural environment probed by fresh honey bee (Hymenoptera: Apidae) pollen', *J. Econ. Entomol.* 110 (2017): 1947–58, doi: 10.1093/jee/tox195

The 'direct' effects of pesticide residues

What are the direct effects of these pesticides on bees? We can define 'direct' effects as clear, evidence-based correlations between increasing pesticide residues and hive mortality. Do surveys show that hives with high levels of pesticide residues are more likely to collapse or die? I'm especially interested in evidence that pesticides have direct effects on bees at a large geographic scale; I'm less interested in reports of isolated events. There is no doubt that pesticides can be lethal if directly applied to bees and individual hives. We occasionally see reports of hive collapse or entire apiary destruction due to the overly enthusiastic efforts of a pest controller. Rather than those isolated and frequently individual events, what is the evidence for pesticides affecting honey bees over large areas such as at the provincial or regional scale? Other, sub-lethal effects that might be mediated through pathogens or altered bee physiology will be discussed below.

The massive, multi-state and multi-country Colony Collapse Disorder observed in the mid-2000s was the motivation for several surveys of the pesticides in beehives of North America. One of these studies is included in the table as US [i]. Were pesticides responsible for this CCD, which resulted in some beekeepers losing 90% of their hives? The symptoms – a rapid loss of adult workers, a lack of dead bees around the hive (possibly indicating they had died quickly elsewhere), and a delayed invasion of hive pests or robber bees[6] – certainly sounded like they could be pesticide-induced. In the US [i] study, Christopher Mullin and a large team of scientists sampled pesticides residues from pollen, hive wax and bees themselves, across 23 states and one Canadian province. Healthy colonies and those displaying CCD symptoms were sampled, as well as bees from a wide variety of agricultural cropping systems and beekeeper operations.

Depressingly, the scientists found a massive number of pesticides and pesticide metabolites (the chemical breakdown products of the pesticides):

We have found 121 different pesticides and metabolites within 887 wax, pollen, bee and associated hive samples. Almost 60% of the 259 wax and 350 pollen samples contained at least one systemic pesticide, and over 47% had both in-hive acaricides fluvalinate and coumaphos, and chlorothalonil, a widely used fungicide. In bee pollen we found chlorothalonil at levels up to 99 ppm [parts per million] and the insecticides aldicarb, carbaryl, chlorpyrifos and imidacloprid, fungicides boscalid, captan and myclobutanil, and herbicide pendimethalin at 1 ppm levels. Almost all comb and foundation wax samples (98%) were contaminated with up to 204 and 94 ppm, respectively, of fluvalinate and coumaphos, and lower amounts of amitraz degradates and chlorothalonil, with an average of 6 pesticide detections per sample and a high of 39. There were fewer pesticides found in adults and brood except for those linked with bee kills by permethrin (20 ppm) and fipronil (3.1 ppm). . . The most frequently found residues were from fluvalinate and coumaphos, followed in order by chlorpyrifos, chlorothalonil, amitraz, pendimethalin, endosulfan, fenpropathrin, esfenvalerate and atrazine. These top ten comprise three in-hive miticides and five insecticidal, one fungicidal and one herbicidal crop protection agents.[7]

Other than two chemicals used for mite control, topping their list of pesticides observed in these North American hives was chlorpyrifos, a widely used organophosphate pesticide that, like many insecticides, acts on the nervous system of insects. The fungicide chlorothalonil, used to control plant pathogens, was also very common.

An article in *Scientific American* sums up the key finding from this and related work at that time: 'Although both the levels and the diversity of chemicals are of concern, none is likely to be the sole smoking gun behind CCD: healthy colonies sometimes have higher levels of some chemicals than colonies suffering from CCD.'[8] Beekeepers in Europe and the United States had long suspected that neonicotinoids were playing a major role in CCD and bee health. Indeed, neonicotinoid residues were periodically found, with the most common being the pesticide imidacloprid. Nevertheless, Mullin and his team wrote: 'Our results do not support sufficient amounts and frequency in pollen of imidacloprid . . . or the less toxic neonicotinoids thiacloprid and acetamiprid to account for impacts on bee health.'[9] The bees from CCD-associated colonies had sub-lethal though high amounts of fluvalinate, amitraz, coumaphos and chlorothalonil.

The other studies presented in Table 1 did not measure hive or colony survival in addition to their pesticide residue analysis. It's unfortunately quite rare to find studies that relate a comprehensive analysis of pesticide contamination with long-term patterns of honey bee colony loss. But, of the studies that have been published, none, to my knowledge, provide evidence that pesticides have a direct role in large-scale colony losses.

For example, a 2010 study in Spain reported that 'a direct relation between pesticide residues found in stored pollen samples and colony losses was not evident'.[10] In a 2009 North American study, researchers examined 61 variables in healthy colonies and in those displaying CCD. The 61 variables included key physiological predictors of bee health, pathogen loads and pesticide levels. The researchers concluded that no factor was likely to be the single cause of CCD. The surviving colonies actually had higher levels of the pesticides that were being used to control mites.[11] At this time, the beekeepers themselves seemed to only occasionally associate colony loss with pesticides. One survey asked North American beekeepers to identify why they thought their colonies died. Only 2.6% thought pesticides were likely to be responsible.[12]

Despite the lack of evidence directly linking pesticide use with CCD or wide-scale colony losses, I think an analysis of chemical residues from within hives is absolutely essential in aiding a country's management and use patterns of pesticides. If we are seeking to minimise the effects of pesticides on bees and other insects, we need to know the specific chemicals that cause harm and the levels to which insects are exposed to them. For example, should New Zealand ban neonicotinoids? I am often asked that question by journalists. In my opinion, in New Zealand we don't currently have enough data to guide that decision.* If just 1–2% of hives here show neonicotinoid residues, while

---

* We know that neonicotinoids are used in New Zealand and can be found in our beehives, though currently we don't know much about the frequency and levels of contamination. The best data we have is from a study of global honey and neonicotinoids by Edward Mitchell and collaborators (*Science* 358 (2017): 109–11). They included just four samples from New Zealand and found all to be contaminated with low levels of neonicotinoid residues. The maximum residue concentration obtained was 0.34ng/g of honey (or 0.34 parts per billion), which corresponds to 0.69% of the Maximum Residue Level (MRL) authorised in the EU. Positive samples were obtained from mānuka honey, which is strange, because mānuka typically occurs well away from agriculture. Perhaps the contamination occurred in one area when the bees were used for pollination, before the hives were moved to harvest mānuka.

other chemicals are more frequently observed in higher concentrations, it may be better for bee health if we focus our efforts on those other pesticides. On the other hand, I'd be a lot more worried and would advocate for change if we, like in France, were to find that 49% of our hives had high concentrations of neonicotinoids. Further, I think we need to link our studies on pesticide residues with our monitoring of long-term hive survival. As our technology improves, we'll be able to find tiny, tiny amounts of an increasing number of synthetic chemicals in beehives. But is this contamination biologically relevant to the health of honey bees?

From these studies, and recent large-scale studies on neonicotinoids that I'll present below, it is clear that no *single* pesticide has been associated with widespread honey bee losses. The next thing to consider is how pesticides might work synergistically. Perhaps combining chemicals is more harmful to bees than the additive effects of each pesticide alone. It's possible that the *indirect* effects of pesticide residues, rather than *direct* effects, substantially influence bee health via sub-lethal or synergistic mechanisms.

Synergistic effects of pesticides on bee health

Many of the studies I've discussed have focussed on the levels and effects of individual pesticides. Over the last decade, however, we've become aware that pesticides and other chemicals can interact to affect the health of bees and other insects. The effects of chemical combinations might be worse than the additional effects of each pesticide alone.

An example comes from the corn-growing regions of Canada. Nadia Tsvetkov, a PhD student at York University, researched the effects of neonicotinoid pesticides on honey bee longevity, behaviour and brain gene expression. In their analysis of pollen and nectar harvested by bees near cornfields, Nadia and her team found a total of 26 different pesticides. The fungicide boscalid caused the highest levels of contamination. (Boscalid is a widely used pesticide for prophylactic fungal control in North America. Three-quarters of ground water sources near farms are contaminated with fungicides, and boscalid is the most common contaminant by far.)[13]* In trials testing the toxicity of boscalid on honey bees, the Canadian researchers found that this fungicide on its own is not harmful. However, when the bees were exposed to neonicotinoid insecticides and these same concentrations of boscalid, the neonicotinoid pesticides became nearly twice as toxic to them.[14]

The magnitude of synergistic effects is important. A two-fold increase due to synergistic effects is significant, but not nearly as significant as a fungicide causing a 16-fold increase in the toxicity of pyrethroid to honey bees.[15]

As we continue to investigate pesticides, we'll discover more synergistic effects. Common synergistic interactions occur between certain fungicides and pesticides that affect the cholinesterase nervous system of insects.[16] For some combinations of pesticides, scientists have shown that synergistic effects may lower a bee's ability to express genes associated with pesticide detoxification (often the 'cytochrome P450' genes). In other situations, we don't know why a fungicide, for example, increases the toxicity of an insecticide. The authors of a 2019 study on the effects of multiple pesticides on bumble bees in the United States concluded that 'both the occurrence and strength of a synergistic interaction with a given insecticide differed depending on the specific fungicide, despite all the fungicides having the same mode of action'.[17]

On a positive note, the bumble bee study and Tsvetkov's study of Canadian corn both highlight that synergistic interactions do not occur between all pesticides. As well as examining boscalid, Tsvetkov and her group looked at how the fungicide linuron interacts with neonicotinoids to influence bee mortality. They found no synergistic effects. The combination of linuron and neonicotinoids was no worse for bees than the neonicotinoid alone. These results are helpful, showing that while pesticides can clearly interact to the detriment of bees, growers and pesticide regulators might be able to select combinations that do not display synergistic increases in toxicity to bees. Ideally, farmers and growers will be supported to develop an integrated pest management plan that has as few effects on bees as possible.

* This study showed up to 2012ng/L of boscalid in US surface water and groundwater samples, though the median contamination level was much lower, at around 20ng/L. Fungicides and herbicides topped the list of most frequently seen pesticides. The most common insecticide contaminant was chlorpyrifos.

Determining the lethality of a pesticide to honey bees is difficult. The toxicity of pesticide residues might depend on the specific environment, other pesticides in the area, and perhaps even the pathogens present. These complications mean that simple estimates of pesticide toxicity, such as an LD_{50}, may have limited ecological value in reality. As we've seen, in worst-case scenarios synergistic interactions can increase a pesticide's toxicity to bees by orders of magnitude. Pesticide regulatory authorities need to be increasingly aware of synergistic interactions and incorporate them into their management. Obviously, it is near impossible to understand and estimate all possible interactions between all chemicals used in agriculture. But a good start would be to examine synergistic interactions between those chemicals that we often see in beehives, as well as the fungicides that are known to interact with many pesticides.

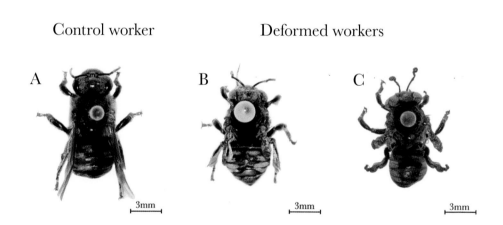

The stingless bee (*Melipona quadrifasciata*) is one of the most important pollinators in the Neotropics. There are many species of stingless bee. The common name of this one, native to Brazil, is 'mandaçaia', which translates to 'beautiful guard'. Above, Bee A has been reared in the absence of pesticides. Bees B and C were reared in the presence of sub-lethal doses of the biopesticides azadirachtin (B) and spinosad (C). B has deformed wings. C has deformed antennae, wings and legs. *Photos: Wagner Barbosa and Raul Guedes*

Sub-lethal effects of pesticides

'What doesn't kill you only makes you stronger.' That saying has always struck me as completely nuts, especially when we consider honey bees and pesticides.

A nice example of the sub-lethal effects of pesticides on honey bees comes from Mickaël Henry's laboratory in Avignon, France. This group wanted to know how sub-lethal exposure to the insecticide thiamethoxam (a neonicotinoid systemic pesticide) influenced honey bee foraging. Specifically, how might consumption of this pesticide alter a bee's ability to find its way home? The researchers first glued little identification tags onto the bees, fed them a low dose of the pesticide, then released the unlucky bees up to 1 kilometre away from their hives. Whether the bees were then able to return home was an indication of 'mortality due to postexposure homing failure'. The results showed homing failure rates of 6–32%. Simulation models using these rates suggested that although hives and colonies of bees were unlikely to collapse, neither would

The honey bee on the left has a radio frequency identification (RFID) chip attached to its back. This chip identifies the bee and allows researchers to know when it leaves or returns to the hive. These chips have been used to understand the foraging behaviour of bees when treated with pesticides such as neonicotinoids. *Photo: Simon Stone, Science Photo Library / Alamy*

they thrive. Hives suffering from sub-lethal effects of thiamethoxam would finish a summer with lower numbers of workers. Their simulations suggested hives might then enter the stressful winter period in a weaker condition.[18]

The work by the French team shows a potential negative, sub-lethal effect of a common pesticide on bee foraging and homing. But we now know that there are, unfortunately, many, many more sub-lethal effects. In 2017, researchers led by Edward Mitchell in Switzerland summarised the known effects of neonicotinoid insecticides on bees. They reviewed studies on how very small doses of pesticides affect the physiology and behaviour of bees. The studies showed that pesticides can reduce a bee's ability to learn, fly, forage and navigate, decrease its sense of smell, cause a loss of postural control, alter its ability to thermoregulate and avoid predators, and decrease its hygienic behaviours. Pesticides might also heighten the bee's susceptibility to pathogen infection. In terms of physiology, we have seen changes in brain function, such as altered nerve firing, gene expression

A scientist with an experimental beehive. Selected bees in the hive have RFID chips attached to them. In this study the bees are subjected to various pesticide doses. They are then released into the experimental beehive and their behaviour is compared to that of healthy bees. The hive has five access tunnels, each with scanners above it to detect the RFID chips. *Photo: Simon Stone, Science Photo Library / Alamy*

patterns, protein content, and immune response. Certainly, pesticide exposure can reduce a bee's lifespan. From this we would expect that the ultimate effect of sub-lethal pesticide exposure is a decrease in the fitness of the hive: reduced colony growth, reproduction, and even later survival, such as in winter.[19]

It's important that we unpack some of the sub-lethal effects of pesticides, with a focus on colony productivity and fitness.

First, let's look at the example of spinosad and bumble bees. Spinosad is a naturally derived biopesticide, certified as being suitable as an organic insecticide in many countries, including New Zealand. A 2005 study showed that bumble bees die if they are exposed to high concentrations of spinosad. The surviving bumble bees displayed a range of sub-lethal effects, including eating less pollen, a delay in larval development, and adults taking longer to forage – all of which resulted in lower colony weights.[20] Spinosad was also shown to have sub-lethal effects on a widespread and important species of stingless bee, causing larvae to develop into deformed workers with compromised walking abilities.[21] Insecticides that don't kill bees certainly don't make them stronger: any insecticide that reduces foraging seems as though it would be likely to reduce colony productivity and weight.

Next, let's consider herbicides. Herbicides such as Roundup target an enzyme system found only in plants, so it shouldn't have major effects on bees or other animals. Recently, however, Roundup has been found to cause sub-lethal effects on the physiology of individual bees. One study found that feeding bees Roundup changed their gut bacteria and made them more susceptible to a bacterial pathogen.[22] Another study found that Roundup caused the degeneration of cellular machinery in the hypopharyngeal glands, which are used to produce royal jelly.[23] Though these findings demonstrate that sub-lethal effects of Roundup are possible, we are yet to see any direct long-term effects of Roundup on bee colony productivity or fitness.

Herbicides have another indirect effect that we should mention. 'Genetically modified herbicide-tolerant' crops resist spraying by herbicides such as Roundup. One disadvantage of this genetic modification is that it leads some farmers to spray Roundup more often. As described in the footnote on page 161, bees are often heavy users of pollen from weeds or non-crop plants. The frequent use of

Roundup has been shown to reduce the abundance of these plants and their pollen, and to change the associated pollinator communities.[24] Genetically modified herbicide-tolerant crops are fortunately not used everywhere (they are banned in New Zealand) and their effects might be mediated by the use of wildflowers or 'Trees for Bees' refuges.

Finally, let's look at the interactive effects of pesticides on bee immunity, pathogen infection and health.

There are certainly studies that have shown that pesticides can modify the immune responses of honey bees. A logical consequence of an altered immune system is greater susceptibility to pathogens. But the results are complicated. One study found that exposure to neonicotinoids increased *Nosema* spore counts, though exposure to fipronil decreased infection by this fungal parasite.[25] A second study showed quite different effects. Exposure to neonicotinoids and organophosphates decreased *Nosema* infections, while acaricides, a herbicide, a pyrethroid and two fungicides appeared to increase *Nosema* infections in the bees.[26] Perhaps the differences between studies were due to differences in bee or pathogen strains. While much of the work examining pesticide and pathogen interactions has involved the microsporidian pathogen *Nosema*, there is good evidence, at least at the level of the individual bee, that pesticides can have synergistic interactions with a wide range of other parasites and pathogens. For example, using miticides to successfully control *Varroa* parasites can significantly increase infections of the deformed wing virus.[27] The neonicotinoid insecticide clothianidin can synergistically interact with the bacterial disease American foulbrood to cause high larval bee mortality.[28]

A major complication, however, is that many of the studies have been at the level of individual bees rather than a hive. Scientists reviewing this literature conclude that the interactions between pesticides and pathogens that we see at the level of individual bees typically do not scale up to colony- or hive-level effects.[29] One of the most comprehensive reviews concludes: 'Pesticide exposure and pathogen infection have not yet been found to interact to affect worker survival under field-realistic scenarios.'[30] Researchers in this area are often as mystified as everyone else, frequently with no explanation of why pesticide and pathogen interactions observed at the individual level do not scale up to affect colony dynamics.

It is abundantly clear that low concentrations of pesticides can have a plethora of sub-lethal effects on bees. But, please, please take note: just because scientists have shown that pesticides *could* have an effect doesn't mean that they do.

A large proportion of these experimental studies have involved a 'no-choice feeding' design. This means that the bees had to feed on the sub-lethal concentrations of pesticides. And often the sub-lethal doses were at a wildly higher concentration than the bees would experience in nature. Mickaël Henry's study on honey bee foragers becoming lost after being given neonicotinoids, for example, has been criticised due to the unrealistic concentrations of the pesticide used. One 2017 review concluded: 'It is extremely unlikely that the findings of Henry et al. are representative of a real-world situation.'[31] Henry's group have since published work demonstrating that while bees might disappear slightly faster when hives are placed near crops with neonicotinoids, the bees compensate for this loss so as to produce an unaltered performance in terms of honey production and population size.[32]

Overall, clearly, we can observe sub-lethal effects if we feed bees a neonicotinoid pesticide, the organic biopesticide spinosad, or the herbicide Roundup. But are hives of bees experiencing these effects in the real world? Does the widespread use of pesticides alter the lifespan and long-term fitness of honey bees and beehives in the real world, under pesticide concentrations and situations that bees actually experience? I'll examine that critical question as we go on to explore neonicotinoid pesticides.

Neonicotinoids: To ban, or not to ban

Neonicotinoids, or neonics, are a group of insecticides that are chemically related to nicotine. They act on the nervous system of insects, stopping the transmission of nerve impulses, which eventually results in the insect's paralysis and death. These synthetic insecticides are much more toxic to insects than to us and other mammals because insects' brains are very different from those of us vertebrates. Neonicotinoids include clothianidin, imidacloprid and thiacloprid, among many others. Imidacloprid is frequently cited as the most widely used insecticide in the world – farmers use it by injecting it into the soil or into trees, or by coating seeds. Seed coating is especially popular because as the plant grows the neonicotinoid is present in low concentrations throughout its tissue (unfortunately including in the bee-harvested pollen and nectar), meaning that multiple foliar sprays are not needed.

Scientists and media often link bee health and colony collapse with neonicotinoids. Let's look at the 2017 *Guardian* article 'The evidence is clear: insecticides kill bees. The industry denials look absurd'. This report sounds definitive and urgent. 'The largest field trials to date offer irrefutable proof. We need a total ban, now, to halt the sabotaging of our own best interests.' This headline alone sounds like something we should write to our local politician about. Let's get rid of these chemicals in New Zealand and everywhere else too! Right? If I were getting all of my science information from stories like this, I'd be convinced. But let's have a closer look at the research on which this article is reporting.

The 'largest field trials to date' did indeed represent a large study. The analysis encompassed 33 sites from Germany, Hungary and the United Kingdom, with plots averaging 63 hectares each. Standardised colonies of honey bees and two species of wild bees (the bumble bee *Bombus terrestris* and solitary bee *Osmia bicornis*) were placed in each site. The sites had the standard crop of winter-sown oilseed rape, which was grown commercially with either seed coatings of one of two neonicotinoids (thiamethoxam or clothianidin) or no seed treatment (control). Fungicides were also used. A total of 258 different analyses were deployed on the data, with the critical unit of measure being 'significant'[*] effects on hive

* We scientists use the term 'significance' differently from the general public. For any experiment or observation that involves taking a sample from a population, there is the possibility that an observed effect would have occurred due to chance or sampling error alone. For example, we might want to know whether beehives exposed to neonicotinoids collapse more often than those without exposure. Our hypothesis is that the mortality rate differs between these groups of beehives. In testing this hypothesis, we would only say the results are 'statistically significant' if we are 95% confident that the average, or sometimes the median, of hive mortality in the two groups are different. Statisticians will say that you can never be 100% confident about anything; but the consensus is that >95% confidence is when we can use the term 'significant' in these sorts of experiments. Statistical analyses are used to calculate this confidence. It is how we are 'sure' of the results of our experiments.

overwintering survival, worker numbers and reproduction.[33]

What did the researchers find? Well, first off, there were no significant effects of neonicotinoids on post-winter survival of honey bees. The scientists had problems in the United Kingdom because *Varroa* killed so many hives there that an analysis was unable to be performed. If we largely ignore statistical significance, the effects on honey bees or native bees were country- and pesticide-specific. In Germany, the use of neonicotinoid pesticides seemed to increase worker and egg cell numbers. In Hungary, one neonicotinoid increased worker numbers, one decreased worker numbers, and egg cell production was reduced for both. In the United Kingdom, one neonicotinoid significantly decreased worker numbers, one had no effects, and egg cell production was increased for one of the insecticides but not the other. There were similarly variable results and a lack of statistical significance for the wild bees. In Germany, the production of bumble bee drones significantly increased, though in the United Kingdom drone production significantly decreased.[34] Of the 258 statistical tests presented, nine showed that neonicotinoids had a significant negative effect on some aspect of bee biology. Seven of the tests showed that neonicotinoid use was, statistically, beneficial.

Does that description and those results paint a 'clear' picture that provides 'irrefutable proof' on how we should manage neonicotinoids? If I stand back and attempt to view this data without any bias, as I hope you might, I can't see the clear patterns that would allow me to make management recommendations. Consistent negative effects of using neonicotinoids were not observed across countries. In Germany, neonicotinoid use even tended to be a positive thing for bees.

Additional large-scale studies have compared bee colonies placed in large neonicotinoid-treated plots of land with colonies in untreated areas. Maj Rundlöf and colleagues in Sweden performed one such study. They did find that a combination of the neonicotinoid clothianidin and the non-systemic pyrethroid ß-cyfluthrin, applied to oilseed rape seeds, reduced wild bee density, solitary bee nesting, and bumble bee colony growth. These results for solitary and bumble bees attracted much of the media attention. But for honey bees:

The insecticide seed treatment had no significant influence on honeybee colony strength. In contrast to the *B. terrestris* colonies, the *A. mellifera* colonies did not differ in strength (number of adult bees) between the treatments after placement at the oilseed rape fields . . . This finding is in line with another field study and previous work suggesting that honeybees are better at detoxifying after neonicotinoid exposure compared to bumblebees.[35]

Different organisations and media outlets reported quite differently on this study. Juergen Keppler, from Bayer CropScience (which produces and sells neonicotinoids), said, 'We are pleased the Research Letter demonstrates again there is no effect of neonicotinoid seed treatments on honey bee colonies in realistic field conditions, consistent with previous published field studies.' David Goulson, a biology professor at the University of Sussex whose research focus is bumble bees, reported, 'At this point in time it is no longer credible to argue that agricultural use of neonicotinoids does not harm wild bees.'[36]

After their first publication and dissemination of their results, the Swedish group extended their study for an additional year to obtain more information on honey bee colony development, swarming, mortality, and pathogens. The authors again found no observable negative effects of neonicotinoids. Instead, 'Clothianidin treatment was associated with an increase in brood, adult bees and *Gilliamella apicola* (beneficial gut symbiont) and a decrease in Aphid lethal paralysis virus and Black queen cell virus – particularly in the second year. The results suggest that at colony level, honeybees are relatively robust to the effects of clothianidin in real-world agricultural landscapes, with moderate, natural disease pressure.'[37] This resilience displayed by honey bees may be because they have effective detoxifying systems, or because they have larger colonies that are able to cope with this environmental stress. Mickaël Henry's group in France believe that honey bee colonies compensate for excess mortality due to neonicotinoid exposure so as to preserve their performance unaltered, both in terms of population size and honey production.[38] In contrast, because solitary bees live alone and bumble bee colonies are small, stress from pesticides might be more damaging. The authors of the Swedish study go on to report, 'There have now been several more or less well-designed field studies that have failed to detect a major impact of field-level neonicotinoid exposure on honeybee colony development, performance and overwintering success.'[39]

All these studies and conclusions are hard to digest. It is no wonder that many beekeepers are confused by the conflicting evidence on neonicotinoids. Here is what I think. These large, multi-year field studies do not tell me that we should be banning neonicotinoids on the basis that they have widespread and consistently negative effects on honey bees. When we look for clear, repeatable evidence that the widespread use of neonicotinoid pesticides alters the lifespan and long-term fitness of honey bees and beehives in the real world, it's just not there. Other scientists have come to exactly the same conclusion regarding pesticides and honey bees: 'Field trials have failed to demonstrate a link between real world exposures and increased mortality.'[40]

I'm not saying that there aren't other potential problems with neonicotinoid pesticides. My focus here is on honey bees, whereas bumble bees, solitary bees and other animals may be affected and should be considered by regulatory

A frame of honey bees from a hive kept beside an oilseed rape field in Sweden. This crop was treated with clothianidin and pyrethroid ß-cyfluthrin. The authors of the study were surprised to find that pesticide treatment was associated with an increase in hive strength and beneficial gut bacteria, and a decrease in infections. *Photo: Maj Rundlöf*

authorities. For example, a 2019 Canadian study showed that songbirds that were intentionally fed neonicotinoids had reduced feeding and accumulation of body mass and fat stores, which led to delayed departure from stopover sites.[41] There is also evidence that neonicotinoids can disrupt aquatic food webs and decrease fishery yields.[42] These and many more studies suggest we should carefully evaluate our use of these synthetic pesticides. We should extend this analysis to evaluate the older chemicals that have been used to replace neonicotinoids where they are banned, as the impact of these older chemicals is often unknown and unquantified.[43] Are countries banning neonicotinoids jumping from the frypan into the fire?

'Save Switzerland from synthetic pesticides'

It is my guess that nearly any chemical used in agricultural environments is likely to have some effect on bees, other insects, and animals in general. After neonicotinoids gained such bad publicity, other insecticides have been developed and marketed as 'bee safe'. For example, the insecticide Sivanto, developed by Bayer CropScience, has a 'bee safe' classification that allows it to be used on flowering crops that have actively foraging bees. Unfortunately, when used in combination with common fungicides Sivanto becomes lethal to bees or has a range of sub-lethal effects, including abnormal behaviour, poor coordination, hyperactivity and/or apathy.[44] (I'm now a little intrigued to see what a hyperactive, apathetic bee looks like.)

These results are not surprising. A honey bee is an intricate physiological and neurological machine, supported by a community of gut bacteria and other symbionts, and vulnerable to a wide variety of parasites and pathogens. The small molecules of many chemicals used in the human world directly affect the cellular functions of honey bees. A fungicide that isn't directly harmful might alter their beneficial gut bacteria and make them more susceptible to pathogens, perhaps also altering their ability to tolerate other pesticides. There is no question that pesticides could be having an array of lethal and sub-lethal effects on bee health that we are yet to discover.

With this knowledge in hand, one logical precautionary approach would be to ban pesticides altogether. Such an idea seems extreme, but that is exactly what is being proposed currently in Switzerland.

The Swiss democracy allows for a national popular vote, or a binding referendum, to be launched when 100,000 public signatures are collected on a proposal within 18 months. On 25 May 2018, a referendum titled 'Save Switzerland from synthetic pesticides' obtained more than 140,000 signatures and was officially lodged with the government. The initiative proposed to entirely ban the use of synthetic pesticides in Swiss agriculture as well as on public land and private gardens. Food produced elsewhere using synthetic pesticides would also be banned from importation into Switzerland. If this referendum is accepted, Swiss farmers and food importers will have 10 years to stop using pesticides and importing pesticide-contaminated food. All of the synthetic pesticides that I've discussed above will be banned, including fungicides, neonicotinoids, and weedkillers such as glyphosate. The referendum organisers believe that Switzerland is able to produce enough food without using these chemicals. Biopesticides will be allowed (there is a little irony here, given that biopesticides such as spinosad can also be damaging to bees). The next-generation pest control approaches using gene silencing or RNAi would likely be acceptable. The referendum leaders cite data showing high levels of pesticide pollution, surveys showing that 90% of soil on Swiss organic farms is contaminated with pesticides, and estimates that 60% of insects in Switzerland are threatened with extinction.[45]

Not everyone, however, has supported the referendum. It is a highly controversial proposal. The National Council of Switzerland's government do not support the initiative or any of the counter proposals. The Swiss People's Party and the Swiss Farmers' Union have both argued that the plan is far too restrictive. The government and farmers believe that the initiative would result in already struggling farms becoming even more economically unviable.

Despite such opposition, the organisers are confident that the referendum will pass and be implemented in late 2020, though they expect a lengthy debate between now and then. Committee member Edward Mitchell, from the University of Neuchâtel, estimates that around a third or a quarter of the Swiss public are concerned about the effects of these synthetic chemicals on bees and biodiversity. 'Most of the public read story after story in the media about how small quantities of synthetic chemicals have devastating effects for human health,' Edward told me. 'Many of the Swiss citizens supporting the referendum are motivated about their own and their family's wellbeing.' He acknowledged that there are crops for which it will be challenging to obtain effective pest control alternatives, such as canola, potatoes and sugar beets. He thinks that some crops aren't suited to grow in Switzerland. 'We are only able to farm them due to synthetic chemicals. Perhaps we should focus instead on the more viable crops able to be grown in Switzerland using organic methods. We can import other foods from countries where these crops are able to be grown sustainably.' The 10-year lead-in time from the legislation passing to being enforced would allow opportunity for new alternative biopesticides or pest control methods to be developed and implemented.

A complete ban on synthetic chemicals is a radical idea, and enforcing it would not be painless. I imagine that the passing of the legislation would result in some multi-generation farmers being forced from their farms and businesses. There will be heartbreak for some, but opportunity for others. From a distant perspective, here in New Zealand, I'm really hoping that the Swiss referendum is successful. It would represent an experiment on a massive scale with an economy that isn't mine. If it is successful, many countries will follow suit – for the possible betterment of bees, biodiversity and human health. And, whatever happens, I'm sure that the biopesticides and environmentally friendly approaches for pest management developed by the Swiss would be useful around the world.

So what should we do about pesticides?

I completely understand the suggested approach of the Swiss group seeking to 'save Switzerland from synthetic pesticides', with their precautionary principle to ban agricultural chemicals. It would drastically reduce the chance of synthetic chemicals affecting bees and could benefit entire insect communities. On the other hand, it could also result in the collapse of many sectors of their agricultural industry.

Should we ban synthetic pesticides too? In New Zealand, an annual survey of beekeepers has been running for several years to help us ascertain reasons for bee colony loss, as discussed in our introduction. These surveys suggest beekeepers in New Zealand do not believe pesticides are a cause of bee mortality here.[46] Beekeepers here believe that queen deaths, *Varroa*, starvation, diseases and other factors such as wasps seem to play a greater role in the health and productivity of our honey bee hives. I think they are probably right.

I often hear people say, 'If only the government would listen and get rid of neonics, our honey bees would be saved.' But I'm not convinced. Research from Sweden and other countries on the widespread effects of neonicotinoids does not convince me that they are a major cause of honey bee mortality. There may be a good case for regulating neonicotinoid use that is driven by the effects of these chemicals on bumble bees, solitary bees, birds or fish – but the case is not there for *honey bees*. In New Zealand we have no recorded or documented bee deaths known to be caused by neonicotinoids.[47] More importantly, how frequently are New Zealand bees exposed to these insecticides? Are neonicotinoids a chemical to which the majority of our bees are exposed, or are they infrequently seen by bees? And if we get rid of this or any pesticide, what will we replace it with?

If I were to target individual pesticides for concern, I'd worry about insecticides such as chlorpyrifos. It is a foliar spray widely used here and in many other countries, though it is banned in others. This insecticide never seems to hit the media but it is known to be lethal to honey bees, as well as having sub-lethal effects on their learning abilities.[48] We commonly see chlorpyrifos in beehives and, in one study, in all samples of Californian breastmilk being given to babies.[49] If we ban neonicotinoids, would we see an increase in the use of chlorpyrifos and other already widely used pesticides? And if we are really trying to 'save the bees' or at least improve bee health, perhaps our efforts would be better placed in regulating and finding alternatives to chlorpyrifos and other pesticides. But, again, before we create a ranking of pesticides that might affect our bees, and before we try to manage their use, first let's see if chlorpyrifos, or neonicotinoids, or whatever other pesticides, are actually widespread in our hives.

While we continue to use synthetic chemicals in agriculture, for pesticide management I think we need three key items of information. Firstly, what are the synthetic chemicals to which bees are being exposed? The amazing answer is that, for many countries, including New Zealand, we currently just don't know. I think frequent, large-scale surveys of the chemicals in pollen and beehives is a must. Secondly, understanding commonly co-occurring pesticides should guide research into and inform us of synergistic effects. It is becoming increasingly clear that these synergistic effects need to be considered in pesticide regulation. Understanding synergistic effects won't be easy and I don't envy regulatory bodies in this goal. Finally, we should combine these pesticide surveys with long-term observations of hives and hive mortality. The longevity and productivity of a hive is the sum of lethal and sub-lethal effects, so this would be an effective way of monitoring both. This data is essential for us to use an evidence-based approach to pesticide management.

Under this scenario of comprehensive monitoring of both pesticides and mortality, honey bees and their pollen could become the canary in the coal mine for insect communities. While we keep using synthetic pesticides, let's sample hives and pollen and let the data do the talking.

A tractor sprays chlorpyrifos over the orange fields in Lindsay, California in 2007. *Photo: ZUMA Press / Alamy*

7. PREDATORS

Ants, small hive beetles, hornets, wasps – and a plethora of other predators

'A day without a friend is like a pot without a single drop of honey left inside,' Winnie-the-Pooh famously said. You can see the extent of Winnie-the-Pooh's fondness for honey by examining the extreme tooth decay inside the bear's preserved skull. Yes, Winnie-the-Pooh was based on a real bear. He was inspired by a female Canadian black bear, named Winnipeg, who was kept at London Zoo. The bear was regularly visited by A.A. Milne and his little boy Christopher Robin. Winnipeg died in 1934. Her preserved skull is missing many of its teeth, apparently as a result of decades' worth of children feeding her honey.[1] It shows the sort of tooth-decay damage we also see in people who are extremely fond of sugar.

Many animals share Winnie-the-Pooh's fondness for honey and honey bees. A beehive is an attractive concentration of nutritious resources. The brood is an abundant protein source and even the adult bees can be a tasty treat of muscle, fat or protein from the haemolymph. The honey is a refined carbohydrate source and one of very few natural sweet treats able to be found anywhere. The wax is also valuable and is food for moths or beetles. Many animals have evolved to specialise in consuming these resources or the bees themselves. Other species attack hives when the opportunity arises.

In this chapter, I'll focus on the predators and pests of honey bee hives that we haven't talked about yet. A range of species can and do affect honey bee hives here in New Zealand and around the world. I'll focus on species that are known to be widespread pests and important causes of hive mortality. The New Zealand colony loss survey has identified social wasps as a leading cause of bee mortality here, with 12% of all colonies lost here to wasp predation.[2] I'll start with these wasps and include a discussion on hornets, which are another predator winging their way to a country near you with alarming swiftness. I'll talk about Argentine ants, small hive beetles, wax moths and many more.

Wasps and hornets

In New Zealand we have two social wasp species that raid honey bee hives: the German wasp (*Vespula germanica*) and the common wasp (*Vespula vulgaris*). They were both introduced from Europe over the last century and both have colonised the length and breadth of the country. Workers of both these wasps raid honey bee hives to steal honey and to kill juvenile and adult bees for food. Wasps have been estimated to cost New Zealand $133 million each year. Much of this cost is due to damage or lost opportunities to the beekeeping industry.[3] These lost opportunities refer to the major problem of wasps collecting resources that bees would otherwise take and convert into honey.

A wasp threatens a honey bee guard at the entrance to a beehive. When abundant, wasps can be a major cause of hive mortality in New Zealand. *Photo: Chris Robins / Alamy*

Wasp control is another major expense. Many parts of New Zealand have low wasp abundances. In these areas bees can successfully defend themselves against wasps, so little is spent on wasp control. In areas of high wasp abundance, however, many beekeepers report that wasps kill about a third of their hives each year. Spring populations of honey bees are often attacked in areas of the North Island. Here, German wasp nests can overwinter, successfully survive the relatively mild winters and grow to sizes containing hundreds of thousands, even millions, of workers. The workers have many voracious little mouths of their brood to feed, which audibly demand food from their older sisters.

Despite the beekeepers' best efforts, hives can be completely destroyed by wasps. Beekeepers describe years in which there are wasp 'plagues'. Entire apiaries are lost. 'We lose more hives to wasps than to *Varroa*,' one beekeeper told me. 'The aftermath of a wasp attack on an apiary is a nightmare.'[4]

A wasp attack on a bee starts with a pounce. The wasp removes the head, wings and other body parts before flying the tasty sections of the carcass back to the wasp nest. If the beehive is strong, the worker bees will defend against an attack by crowding the entrance, swarming over the wasps and smothering them. Unfortunately, wasps often seem to focus their attacks on weaker hives that can be overcome by sheer numbers. They will take any opportunity to slip into the hive through a crack or gap in a rotting box. An all-out assault then begins.

The most effective method of wasp control we have in New Zealand comes via a toxic bait called Vespex, which is attractive to wasps but not honey bees. Bees ignore Vespex because the bait matrix is protein-based with few carbohydrates. The low concentrations of the insecticide fipronil mean that a wasp will take the bait to its nest and distribute it to larvae and the rest of the wasps, prior to any toxic effects taking hold. Vespex was developed here largely due to the need for wasp control in native forest ecosystems, but perhaps the biggest user population is now beekeepers. Future options for control range from biological control to the use of genetic modification tools.[5]

Globally, however, there are other winged killers that keep beekeepers awake at night.

There are two species of hornet that I think we should worry about. First is the Asian hornet, also known as the yellow-legged hornet (*Vespa velutina*). Hornets are very similar to wasps: they represent a large predator force that can arrive en masse to rob and frequently kill entire hives. The Asian hornet is native to Southeast Asia, though has been introduced into Europe. It is thought to have arrived in France on a container of pottery in 2004.[6] Despite an extremely low level of genetic diversity, this hornet seems to be doing well and continues its spread throughout Europe, with a distribution that now includes Italy, Spain and the United Kingdom. It hunts on the wing, chasing down flying insects that can include dragonflies and, unfortunately, honey bees. When a colony of Asian hornets finds an apiary, they typically then specialise in honey bees. A foraging hornet hovers near a hive, scanning for returning workers. Once it has captured a bee, the hornet flies off to its nest. In some areas of Europe this predation has resulted in the loss of 30–80% of hives, with surviving honey bee colonies having their worker numbers halved.[7]

The other species of bee-killing hornet is one scary beast. The Asian giant hornet (*Vespa mandarinia*), also known as the yak-killer hornet, has been established

Asian hornets (*Vespa velutina*) hover, waiting to catch and kill honey bees returning to the hive. The bees are attempting to guard the entrance.
Photo: Simon Stone / Alamy

in North America. May Berenbaum, Head of Entomology at the University of Illinois at Urbana-Champaign, told the *New York Times*, 'You want to talk about beepocalypse, they are sworn enemies of honey bees. I would say a bee's worst nightmare. Probably the worst nightmare of a lot of people, too.'[8] Laura Lavine, from Washington State University's entomology department, agrees. 'I think there's a need for panic. I've heard anecdotes of beekeepers in Asia standing around with badminton rackets, smacking the hornets to the ground and stomping them. I know that sounds totally insane, but that's what it can come to.'[9] Like the Asian hornet, the Asian giant hornet is thought to have arrived in a shipping container, and is still spreading.

As these hornets spread throughout Europe and North America, the chances of their being transported elsewhere around the world increases. All it takes is for a single mated queen hornet to decide that a large shipping container (or an item going into it) is an ideal place to hide and overwinter. And *voilà*, the hornets soon have a new home and new bees to eat.

Argentine ants

Originally from South America, Argentine ants (*Linepithema humile*) have become an invasive species that is now common in countries with Mediterranean and subtropical climates. They have been described as 'the Genghis Khan of the ant world', due to their decimation of most native ant communities they invade.

The Argentine ant is brown, about 2 millimetres long, and otherwise nondescript. One distinguishing feature, however, is its abundance. If you have Argentine ants, you are likely to have a lot of Argentine ants. Their high abundance is in large part due to their lack of aggression towards other Argentine ants. In most ant species, different colonies hate each other. Neighbouring colonies of the same species are in constant warfare. But not so with Argentine ants; colonies in the same area will pitch in with food harvesting or battles with other species. Nests scattered over wide areas are referred to as 'supercolonies'. One vast supercolony of Argentine ants in Europe is reported to stretch 6000 kilometres along the Mediterranean coast.[10] Another supercolony extends over

900 kilometres along the coast of California in the United States. Ants taken from these European and Californian colonies are friendly and cooperative.[11] Individually, these ants aren't especially strong or powerful, with one-on-one battles typically resulting in the death of the ant. But what Argentine ants lack in individual strength they make up for with overwhelming abundance.

In New Zealand we see Argentine ants attack and kill bees, especially in the far north, where it is warm and dry. Like wasps, these ants often focus on an individual hive in an apiary. Massive numbers of ants will flood in to the hive. Ant queens and ant brood are frequently observed in hives, which tells us that beehives, offering food and shelter, can be used as nest sites for ants. They then steal the honey and eat the brood. The adult bees try to ward off the tiny intruders, buzzing their wings in an attempt to blow them away. The bees are clearly stressed and do their utmost to drive away the ants, with little or no success.

Argentine ants (*Linepithema humile*) in a beehive. The ants enter through cracks in the hive and steal honey and attack bee brood. Hives collapse under heavy attack, or the bees abscond, leaving behind honey and brood. *Photo: Phil Lester*

In 2018 Argentine ants were estimated to cause approximately 0.2% of all hives lost in New Zealand,[12] indicating that they're not a massive problem for beekeepers here. In countries where the ants are common, beekeepers will tell you that they have to abandon entire apiary sites due to constant raids. We've also found that the ants host and carry a range of viruses previously thought to be limited to honey bees, including the deformed wing virus and Kashmir bee virus.[13] Other ant species that co-occur with honey bees carry viruses commonly seen in honey bees.[14] How much of a role the ants have in the transmission of these viruses to honey bees remains to be seen.

Currently, the only ant control options we have are chemical pesticides. As with wasps, protein-based baits are available that are attractive to ants but not to bees, and which contain a low dose of pesticide that ants take back to the nest and distribute to the colony. Some beekeepers are currently dousing apiary sites with contact insecticides such as the termite-killer Termidor, which contains a high concentration of fipronil. The high concentration kills insects on contact and then creates a barrier of residues that ants, and any other sensible insect, won't crawl over. But these residues, and the risk of them washing into waterways, make Termidor an undesirable method of ant control around beehives.

Other ant species can be a problem for honey bees. The red imported fire ant (*Solenopsis invicta*) was reportedly introduced to California by hitchhiking on beehives that were being transported for pollination purposes.[15] (I think Argentine ants are moved around New Zealand in similar ways.) Beekeepers in Texas report a variety of ant species, including fire ants, robbing pollen and sugar resources from hives, attacking and eating brood, scavenging dead adult bees, and cohabiting with bees in their hives.[16] Hives that have collapsed or are close to it are attacked more often. These ants can be helpful in one way, however – occasionally they're reported to control wax moth on stored honey comb.[17]

Small hive beetles

'Unfortunately, the insect came to stay and to this day continues to affect beekeepers throughout Yucatán,' said Mexican beekeeper Marcos Rafael

Chan Cahuich in 2019. 'When the beetle enters the hive, it alters the honey by fermenting it, but the worst thing is that it devours the larvae, until the hive dies or emigrates.'[18] The beetle that Chan Cahuich is referring to is the small hive beetle (*Aethina tumida*).

The small hive beetle is native to the sub-Saharan countries of Africa. Since 1996, however, this species has been spreading around the world. The most recent recorded invasion was in Brazil and South Korea in 2016, though it has been in Australia since 2002.[19] It is a small brown beetle 5–7 millimetres long and is rarely a pest in its native range. But its impact on bees in its invaded range can be devastating. It is a parasite and predator of both honey bees and bumble bees. Its larvae damage the hives by feeding on pollen, honey and bee brood. Combs are destroyed. A slimy material covers the honey, which is partly a result of the honey fermenting and partly derived from beetle larvae. This slime contains a yeast that appears to attract even more adult beetles. Entire beehives can be destroyed. Small hive beetles also appear to vector the bacterial disease American foulbrood (*Paenibacillus larvae*) and viruses including DWV and SBV.[20]

Larva of the small hive beetle (*Aethina tumida*). These larvae feed on the pollen, honey and bee brood. *Photo: USGS*

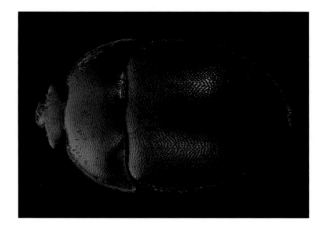

An adult small hive beetle. Adult beetles are good fliers, thought able to fly up to 10km to find a new beehive. *Photo: USGS*

The life cycle of the beetle begins with an adult entering a beehive. It seems likely that the beetles use chemical mimicry to fool bees into accepting them into their hive. The bees even feed the adult beetles.[21] Even if it is attacked, an adult beetle is protected by its defensive body form. It has sclerotised shields on its body and can draw its legs under itself and lower its head, a position called the 'turtle defence posture'. In this position the beetle is very difficult for a bee to grasp or sting. The female beetle will lay eggs in the cracks and crevices in a hive, or directly on pollen and brood combs. One female might lay 1000–2000 eggs in her lifetime. These eggs hatch and feed on the hive, maturing to larvae about 11 millimetres in length over a 7- to 16-day period, depending on the temperature. The mature, 'wandering' larvae then exit the hive and burrow into the soil, where they pupate. The larvae might walk more than 200 metres to find soil suitable for pupation. Adult beetles emerge three to six weeks later. These adults can live for over a year and fly up to 10 kilometres to find host hives.[22] The beetles are able to withstand cold conditions, overwintering successfully in hives by clustering. They seem likely to tolerate the entire range of temperatures and conditions experienced by bees around the world.

Bees aren't defenceless against small hive beetles. The first line of defence is the guard bees at the hive entrance. The small hive entrance, copious use of propolis, and high aggression by Africanised honey bees helps explain why these bees are less afflicted by small hive beetles than western honey bees are.

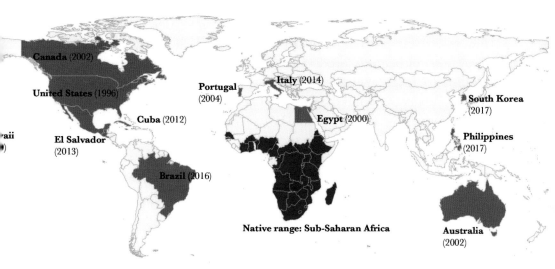

Global distribution of the small hive beetle, with dates of first discoveries

Bees can attack and injure beetles despite their turtle defence, even carrying off the beetles' larvae and dropping them away from the hive. Another anti-beetle defence by bees is to build propolis prisons, into which they corral beetles. Some bees stand guard over the corralled beetles, attacking them if they move, while others build the propolis prison.[23] Bizarrely, the adult imprisoned beetles won't starve, because they get food from their guards. The beetle taps the bee's mouthparts with its antennae, seemingly duping the bee into regurgitating food for it. Often, the guard bee seems to recognise it has been fooled and attacks the beetle soon after the feeding.[24] Their final defence against small hive beetles is abandonment of the hive. This absconding and swarming in response to beetles is most commonly seen in Africanised honey bees and seems to occur when some damage threshold is met.

The small hive beetle is continuing to spread around the world. It now has a foothold on all continents and will continue its march. How do we stop it?

Prevention is easier than cure. The small hive beetle was previously thought to be introduced primarily through shipments of bees from one country to another. Supporting this hypothesis, shipments of live bees from Australia seem to have been responsible for an incursion of these beetles into Alberta, Canada, in 2006. However, recent research has produced strong evidence that shipments and the importation of beeswax products is a primary cause of beetle invasions. Nine of the twelve invasion events examined seem likely to be due to the shipment of beeswax products around the world.[25] These shipments are typically by boat, so a good preventative measure would be to closely monitor ports. Strict regulations on the importation of honey bee products or equipment, like the regulations we have in New Zealand, will be beneficial too.

If small hive beetles slip past the border, speedy detection is essential if there is to be any chance of eradication. Successful eradications have occurred in Perth, Australia and Sicily, Italy. These eradications typically involve sealing the infected hives, destroying bees and beetles using pesticides or by burning, and subsequent monitoring. It's highly recommended that sentinel hives be placed near points of entry. If eradication is not possible, containment to an area can be considered using traps, soil drenches, or medication to hives. It's important that countries have those tools available in advance.[26] If containment is the only option, however, that has major ramifications for beekeeping (such as for an export industry that moves live shipments of bees internationally).

Wax moth

There are two species of wax moth that inhabit hives all over the world: the greater wax moth (*Galleria mellonella*), and the lesser wax moth (*Achroia grisella*).

The greater wax moth is considered the more damaging pest in honey bee hives. It has been described as a ubiquitous pest found everywhere beekeeping is practised. It is widespread in Europe, North America and New Zealand, although there are countries from which it has not been described.[27] Larval feeding is the problem: the greater wax moth consumes nearly everything it finds in the hive. The larvae will eat honey, pollen, wax, honey bee pupal skins, and even other wax moth larvae if they're hungry enough. Their feeding damages the brood and honey frames as they burrow through the hexagonal cells. Their tunnelling leaves masses of silk webs with insect frass (faeces) on the walls and over the comb surface. If the silking becomes excessive the hive can experience galleriasis, meaning that the silk entangles and traps adult bees attempting to emerge from their pupal cells. Scientists have found spores of American foulbrood in moth frass, as well as pathogens such as honey bee viruses in the moth larvae themselves, although the role of the moths in the spread of these diseases is unknown. They even host the bacteria that cause severe pneumonia or Legionnaires' disease in humans. Greater wax moths have been associated with honey bee colony loss. Already weak colonies appear susceptible to infestations and, if a moth infestation becomes sufficiently damaging, the entire bee colony may abscond.[28] The greater wax moth is severely damaging in sub-tropical and tropical regions.

Lesser wax moths are less of a problem. They are smaller and less common in beehives, though they still have a global distribution that includes New Zealand. They feed in a similar way to the greater wax moth and similarly prefer a warmer climate. Extreme infestations of the lesser wax moth can cause 'bald brood' within hives.[29] This syndrome or disease is due to pupal cells being uncapped before the bee has finished developing. The resulting adults appear bald. Bald brood is associated with other pests in hives, including the *Tropilaelaps* mite, a genus of mite which many believe would worsen existing problems for honey bees if they were to spread outside their current range in Asia.[30]

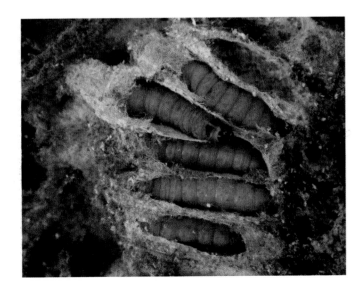

Greater wax moth larvae (*Galleria mellonella*) with their webbing inside a beehive.
Photo: Justine Peacock / Phil Bendle Collection

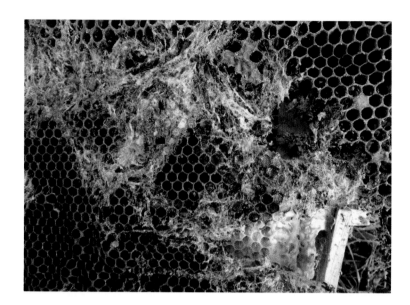

Damage inside a beehive due to greater wax moth larvae. *Photo: Justine Peacock / Phil Bendle Collection*

Usually both wax moth species can be controlled through good hygiene and sanitation. A strong colony with plenty of resources will control wax moths themselves. Beekeepers should destroy infected frames and store equipment or supplies in a way that prevents infestation. Infected material can be frozen or fumigated. As discussed above, some beekeepers in fire ant–infested regions leave frames out to be 'cleaned' by these invasive predators.

It's easy to think of wax moths as disgusting and depressing, especially to beekeepers. However, they have another side that could help humanity. Some observant scientists have discovered that wax moths can eat and digest plastic, or at least the polyethylene and polypropylene that represent >90% of total plastic production. These moths digest the plastic and convert it to ethylene glycol.[31] It turns out that the moths carry mutualistic gut bacteria that are good at digesting complex molecules of the sort you find in beehives, such as wax. Researchers have extracted these bacteria and can grow them on a diet of plastic for over a year.[32] Wax moths and their bacteria might turn out to be extraordinarily useful, given the massive global problem of plastic pollution.

There are other moths that forage in beehives. You can sometimes see the Indian meal moths (*Plodia interpunctella*) or Mediterranean flour moths (*Anagasta kuehniella*) inhabiting hives or hive products. The most fantastic moth to find in a hive is the death's-head hawkmoth (*Acherontia atropos*). This species is famously featured in the Hannibal Lecter film *The Silence of the Lambs*. The adult moths, which can grow up to 12 centimetres long, have a pattern on their back that looks like a human skull. The moth squeaks when predators approach. Scientists have observed that even decapitated moths will squeak when prodded.[33] The squeak seems to mimic the piping sound that queen bees make, which can tell the worker bees to stop moving. In its native range of Africa and the Mediterranean regions, this moth will stroll into beehives and drink honey. This level of nonchalance is achieved by the moth producing a chemical that mimics the scent of bees, so the bees aren't provoked to defend their honey stocks. Despite its startling appearance in popular culture, including in Bram Stoker's *Dracula*, the moth is at worst a minor pest in beehives.

Tracheal mites

How would you like having dozens of little parasites crawling around in your lungs? What about if these parasites mated and bred inside your warm, moist organs, penetrating the thin skin of the lung material in order to suck your blood? The tracheal mites that inhabit honey bees nicely fit that description.

Tracheal mites (*Acarapis woodi*) are about 0.15 millimetres long – too small to see with the naked eye. They live inside the tracheal breathing tubes or internal air sacs of honey. It is here that they feed on the bee's haemolymph, and mate and reproduce. Younger adult bees, up to three days old, seem most attractive to the adult female mites, which disperse into the bees at night. A bee can become so infested that its tracheal tubes become darkened, appearing scarred and brittle, with a secondary film or coating on heavily infected bees.[34] Breathing becomes hard for the bee. It loses haemolymph, and microorganisms can gain entry to the bee through mite feeding, all of which can culminate in the bee's death. Tracheal mite infestations shorten bee lives and honey production by about 20%. Of most concern is that when more than 30–40% of a colony is infested with tracheal mites, the colony is more likely to die in winter.[35]

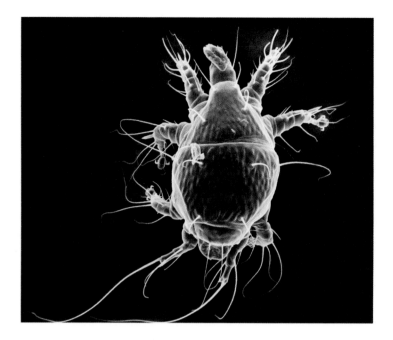

A male tracheal mite (*Acarapis woodi*). These are tiny mites. Females are about 0.15mm in length, males are slightly smaller, and both are too small to see without magnification. *Photo: Erbe Pooley / USDA*

Tracheal mites were first described in 1921 in Great Britain. At this stage they were thought to be associated with the Isle of Wight disease. But it is likely that other factors were involved instead, with one author concluding that 'evidence shows that *A. woodi* was not obviously pathogenic and certainly could not have been causing the observed sickness which was considered to be the IOW disease'.[36] There is much more evidence that tracheal mites were a substantial problem after their introduction into North America. Tracheal mites were first found in Mexico in 1980 and had spread into Texas by 1984. The movement of hives for pollination purposes or the sale of packaged bees then facilitated their movement throughout the country, whereupon they played a major role in colony loss. After reaching Pennsylvania, for example, beekeepers with tracheal mites lost 31% of their overwintering colonies, compared with just 11% losses experienced by neighbours without these mites.[37] Very similar colony losses were observed in the Pacific Northwest.

For quite some time after its introduction and spread throughout North America, honey bee tracheal mites were considered 'second only to *Varroa destructor* as an economically important parasite of adult honey bees', as authors of a 2005 study wrote.[38] It is still widespread but is considered less of a problem now, probably because of resistance development in honey bees. Bees can develop mite resistance through autogrooming or grooming from other bees. When a bee autogrooms, it moves its midlegs over the dorsal areas of its thorax, which removes migrating tracheal mites. Researchers confirmed that this is effective after experiments involving the removal or amputation of these bee legs.[39] The autogrooming behaviour is an inherited trait and is common in 'Russian bees', also known as 'Buckfast bees', although other strains appear to have developed similar resistance. Honey bees also elicit grooming from other bees (allogrooming) using a grooming dance – for a period of about nine seconds, the bee stands stationary and vibrates her whole body from side to side.[40] These dancing bees concurrently autogroom with their midlegs, all of which increases the chance of attendance by another bee. A higher mite infection results in more dancing and allogrooming, and again, this behaviour is genetically inherited.[41]

The mite has spread nearly everywhere there are honey bees, with the exceptions of Australia and New Zealand. In Japan, tracheal mites were first observed in 2010. There, two species of *Apis* bees occur: the European honey bee *Apis mellifera* and the Japanese honey bee *Apis cerana japonica*. The mite spread rapidly across the country, causing the collapse of Japanese honey bee colonies while leaving European bees largely unaffected. Behavioural studies indicated that the Japanese honey bees were much less efficient in their autogrooming than the European honey bees.[42]

How can we control tracheal mites? Using strains of bees that are resistant to mites, primarily through their grooming behaviours, is key. The first strain selection for resistance took place after the Isle of Wight disease, by Karl Kehrle, also known as Brother Adam, of Buckfast Abbey, Devon, England. In 1916 all but 16 of the 46 hives at the Abby died as a result of the tracheal mite's arrival. From the surviving hives Brother Adam created the Buckfast strain. He was known as a talented beekeeper. One account of his work describes him talking to and stroking his bees. 'He brought to the hives a calmness that, according to those who saw him at work, the sensitive bees responded to.'[43] Today, Buckfast bees are known for their productivity, calmness and disease resistance, and can be purchased all over the world.

Other methods of tracheal mite control include the use of chemicals such as menthol, formic acid or miticides.

European foulbrood

Of all the diseases that make your beehives stink, it's American foulbrood that gets the most press. European foulbrood (EFB) typically takes second place. Nevertheless, EFB is a widely distributed pathogen that leads to disease and death in larvae, and in severe cases the complete collapse of honey bee hives.

EFB is caused by the bacterium *Melissococcus plutonius*. It is a disease that typically kills the larvae when they are four or five days old. It has two distinctive

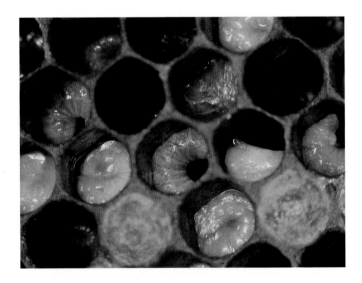

Honey bee brood displaying an EFB infection.
Photo: Eva Forsgren

features: the colour of the larvae, and their position in the cell. Infected larvae change from a pearly white colour to yellow, then brown, and finally a greyish black as they die and decompose. When a high proportion of the larvae are infected, the hive and the decomposing larvae give off a sour or foul smell. The larvae are twisted around the cell walls or sometimes are stretched out lengthways.[44]

The infection cycle involves nurse bees feeding the larvae infected food. An EFB infection can begin with fewer than 100 bacterial cells. Resulting infections are frequently lethal but aren't necessarily or immediately so. The larvae might die and be removed from the colony. It might also reach pupation, which isn't a positive event for the colony, because these larvae will have voided their bacteria-laden intestinal contents into the cell before pupating. Their cells then become primed for future infections. Occasionally the infected larvae survive pupation and may become undersized or even normal-sized adults. Apparently healthy and asymptomatic hives may still contain adults that carry EFB.

There are differences between strains of EFB, just as there are with American foulbrood, *Crithidia* and DWV. The effects of EFB are influenced by both the genetic strain of the host and the strain of the pathogen.[45] Honey can also be infected. This disease often spreads via the beekeeper via contaminated equipment or the transfer of frames between colonies. The bacteria can survive for months on infected equipment. Infections can also occur when bees from healthy hives rob diseased or weakened hives.

EFB has a wide distribution and occurs nearly everywhere there are honey bees, with the exception of New Zealand and a few other countries. In England and Wales, EFB has replaced American foulbrood as the most widely observed bacterial disease observed in hives. There, EFB is sometimes referred to as a 'stress disease'.[46] Hives showing EFB symptoms might recover and show no symptoms, but if the colony is under stress the disease will return. EFB can be controlled through the destruction and burning of hives, which is legislated in countries or regions such as Western Australia where EFB is absent. In some countries antibiotics and shook-swarming are used.

A lot of things eat bees

Many other predators, parasites and pathogens attack and eat bees. I've not discussed all of them – instead, I've tried to limit the discussion to species that are known to be widespread and major causes of bee mortality.

There are other species peeking over the horizon as potential major pests of bees. *Tropilaelaps* mites, for example, are originally from Asian giant honey bees (*Apis dorsata*). Some consider these mites a serious emerging threat to European honey bees. They transmit viruses including DWV and ABPV, which are associated with deformities and reduced bee longevity.[47,48] They have now expanded their range into areas of East Asia including South Korea and China. One of the key issues with this mite, just as for *Varroa*, is that it hasn't co-evolved with the European honey bee, which places the bee at considerable risk.[49] These mites are now classified as 'notifiable pests' in countries such as Austria and Australia. This and other emerging parasites or pathogens that have co-evolved with species other than European honey bees are especially worrying for honey bee health.

Winnie-the-Pooh is the least of our concerns.

8. THE FUTURE

A lot of things affect bee health –
and sometimes populations collapse

Honey bees have a lot to deal with. They are afflicted with disease and parasites and eaten by predators, and their populations sometimes collapse. We've known all this for over a thousand years. Aristotle talked about disease in honey bees, but it was George Fleming who reviewed animal plagues and described what was probably the first recorded widespread 'mortality of bees', which took place in Ireland in 950. Collapses of honey bee populations have continued to dot the timeline of our recorded history. Perhaps we've seen even more colony collapse events during the last 100 years, due to the exponential increase of human populations, combined with the global movement and culture of bees. Some mass mortality events gain notoriety, such as the Colony Collapse Disorder (CCD) in North America. Other crashes in honey bee populations receive less attention, such as the widespread loss of honey bees ongoing in Russia as I write this chapter.

What causes widespread collapse in honey bee populations? The bottom line is we don't know their specific cause. A massive amount of money and effort went into attempts to understand CCD in North America and Europe. But, despite work from some of the world's leading bee researchers, we have no smoking gun. Scientists have argued about the role of one factor or another for CCD, or the loss of 90% of bee colonies from the Isle of Wight disease in the early 1900s, without any definitive conclusion or consensus. Unproven theories for colony collapse range from cellphone towers[1] to neonicotinoid pesticides.[2] It seems unlikely that either of these factors was responsible for the Isle of Wight disease, or any of the many, many other mass bee mortality events before then. About all we can conclude is that collapses are often associated with environmental stress. An extended period of unusually cold and wet weather is often seen prior to colony collapse. Sometimes collapses are associated with a pathogen (or strain of pathogen) or parasite that has recently invaded and established. The parasites I've discussed, including *Varroa* mites and tracheal mites, can and do kill honey bee hives. Pathogens such as the deformed wing virus, American foulbrood and *Nosema ceranae* can also be devastating. Their effects can be worsened by environmental stress. The misuse of pesticides also occasionally kills populations of bees.

I'm confident, unfortunately, that we will see more widespread collapse of honey bee populations over the next few decades. There is little we can do to stop these events in the absence of a single known cause. Events with unknown mechanisms are difficult to manage. When pathogens and parasites are involved, our beekeeping methods almost certainly exacerbate the spread and impact of these devastating events. We use apiaries that have unnaturally large numbers of hives and bees. Bees drift between hives and consequently parasites and pathogens do too. Bees from a strong colony rob disease-weakened hives, or happily find a home in a new hive when a beekeeper transfers frames between hives and use pathogen-infected equipment in multiple colonies. Importantly, we also transport bee colonies widely around the countryside for pollination events. Moving around and congregating 30 billion or two-thirds of America's bees by truck into California's almonds and then sending them home again[3] is a great way to spread and share parasites and disease around a nation.

But what can we do to improve bee health?

Big drivers of bee hive losses: 'Queen problems' and starvation

Each year a survey of honey bee colony losses is undertaken in many countries around the globe. Its primary purpose is to assess overwintering colony loss, winter being a key period for honey bee mortality. What do the surveys find? Rates of mortality differ substantially between countries. For example, data from colony losses over the 2016 period range from Norway's 7.7% to Germany's 44.5%. New Zealand has rates at the low end of that spectrum.

Table 2: Overwinter colony loss estimates, 2016

Country*	% loss from queen problems	% overall winter loss rate from all reasons
Norway	3.7	7.7
New Zealand	2.9	9.8
Northern Ireland	6.3	10.0
Algeria	1.8	10.8
Sweden	3.3	15.2
France	4.1	19.5
Belgium	3.9	23.4
Serbia	6.6	24.1
Mexico	11.6	25.3
US	4.3	26.9
Spain	8.6	27.6
Germany	11.5	44.5
Overall	5.1	20.9

*Data for all countries are from Brodschneider et al.[4] except for New Zealand (Brown et al.)[5] and the US (Kulhanek et al.).[6] There is no data available for several countries, including Australia.

Countries are arranged from lowest to highest overall winter rate of loss. Countries with the five highest and five lowest rates are shown, as well as other countries of interest. The percent loss directly attributable to 'queen problems' is also shown. Queen problems are typically the largest cause of mortality each year; in New Zealand in 2016, just under a third of all reported losses were attributed to this. It's important to take this sort of data with a grain of salt or two, as surveys are all run slightly differently and with varying rates of response by beekeepers.

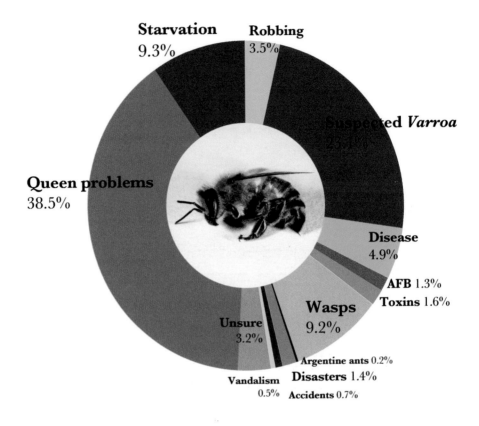

Starvation 9.3%
Robbing 3.5%
Suspected *Varroa* 23.1%
Queen problems 38.5%
Disease 4.9%
AFB 1.3%
Toxins 1.6%
Unsure 3.2%
Wasps 9.2%
Argentine ants 0.2%
Vandalism 0.5%
Disasters 1.4%
Accidents 0.7%

Sources of overwintering honey bee colony mortality in New Zealand in 2018.
Source: Brown and Robertson, Report on the 2018 New Zealand colony loss survey

The beekeepers undertaking these surveys always report that 'queen problems' or 'queen quality' are a major cause of overwintering colony losses. Why?

There is only one queen in the colony. If she dies the colony is in trouble, especially over winter when there are few or no eggs that the workers could develop into a new queen. It is important to remember that queens do naturally die in colonies. Beekeepers can also contribute to an unnatural death when they move hives or frames around. Queens can live for four or more years, but reports of a substantially reduced longevity, with queens living for less than one year, are becoming common.[7] Relatively young queens are more likely to survive overwintering than older queens.[8] Reasons for variable longevity or 'queen quality' are frequently debated. There is good evidence that multiple matings are essential, with one study finding that there appears to be a threshold of least seven males. Colonies with queens that have mated with fewer than seven males were nearly three times more likely to die at less than one year old, compared with queens that mated with more than seven males.[9] The genetic diversity and quality of these drones is important. Older queens slow in their brood production and can also run out of stored sperm and lay unfertilised eggs. Although a newly mated queen will return to her hive with 10–20μL of semen that contains around 100 million spermatozoa, most of this semen is discharged, leaving only 3–5% of each drone's sperm to be stored in the queen's spermatheca.[10] The longevity of drones' semen can also be strongly affected by pesticides, including the miticide chemicals used to control *Varroa*.[11] A queen's longevity can also be reduced by pathogens.[12]

Another common problem seen in many of the surveys is starvation. There is global consensus that damage to bees' natural habitats, resulting in a loss of flowers and resources, is a long-term driver of bee declines.[13] Starvation effects are increasing in New Zealand, even though our country is often associated with the gold-rush of mānuka honey. Beekeepers enthusiastic to profit from the high value of mānuka may overstock regions. Maggie Olsen, from Mānuka Farming New Zealand, recently said beekeepers are in intense competition with one another and people are no longer respecting the traditional rules. 'There's about a million hives in New Zealand now but in saying that there's a lot of boundary riding. Bees would usually be spaced about 3 kilometres apart, now people will put hives anywhere that has space for them.' All this is 'just plain stupid'.[14] Also, bees evolved in landscapes with many diverse plants, but in a modern agricultural environment they typically have a monotonous diet. In previous chapters we've seen how pollen or nectar from plants such as thyme or sunflowers can combat pathogen infections. Honey bees appear to be able to 'self-medicate' after infection, but they need the right resources from the right plants to be able to do so.[15]

In New Zealand, programmes such as Trees for Bees can help provide more pollen and nectar resources. In most countries, however, there is a dearth of bee resources despite similar programmes. As Dave Goulson writes, 'The take-up of schemes to boost pollinators remains low in most countries, perhaps reflecting a lack of understanding of the economic and environmental benefits or a lack of familiarity with implementation of such measures.'[16] Insect communities as a whole would benefit from habitat conservation, given that habitat loss is consistently rated as the major cause of pollinator and insect declines. We are currently experiencing an 'insect apocalypse', because we have 'failed to appreciate the full scale and pace of environmental degradation caused by human activities in the Anthropocene', says Goulson.[17] If we are to conserve insect communities, we need to increase bee resources at a large scale.

A beekeeper with a frame of mānuka honey. Mānuka honey has exploded in value in recent years. *Photo: Boaz Rottem / Alamy*

Other causes of colony loss: *Varroa*, disease and everything else

'Queen problems' and starvation made up just under half of all the estimated causes of overwintering colony losses in the 2018 New Zealand survey.[18] What's left? The biggest remaining part of the pie is mortality from *Varroa*. This pattern of honey bee mortality is echoed around the world. From a study on predictors of colony strength and survival in Britain: 'Our results suggest that beekeepers in England and Wales should concentrate on *Varroa* control, queen maintenance, and timely feeding to reduce colony losses.'[19] Similarly, North American textbooks emphasise the importance of large colonies with young queens, and making sure they have access to high-quality food, a good climatic environment and low levels of parasitic mites and disease.[20]

Wasp predation, disease, American foulbrood, and other factors such as Argentine ants and vandalism account for the rest of the mortality pie. The impact of wasps and Argentine ants varies between regions, given that these

There are dozens of colonies in this apiary. Such an arrangement of hives is useful for beekeepers. Epidemiological theory predicts, however, that disease will be more prevalent where hosts are highly concentrated. *Photo: Saša Lalić / Alamy*

insects aren't uniformly distributed or abundant. Control options are available for both, typically involving pesticides. Perhaps of more concern to beekeepers are the novel or emerging species that are driving colony losses in Asia, such as the *Tropilaelaps* mite, which has co-evolved with the giant Asian honey bee.[21] The yellow-legged hornet (*Vespa velutina*) is also a significant 'alien driver of honey bee colony losses' as it spreads through Europe and around the world.[22] We have few control options available for hornets or *Tropilaelaps* mites at this time.

It's important to remember that colony loss surveys are based on beekeepers' perceptions. I know many beekeepers who know a lot about their bees. But beekeepers around the world do have varying abilities in diagnosing problems with beehives. Pesticides could be playing a hidden role in many of these categories of bee mortality, including 'queen problems'. Even though we know that, as recent reviews suggest, 'field trials have failed to demonstrate a link between real world exposures and increased mortality of hives', we cannot exclude the possibility of pesticides influencing colony mortality in some way.

How would you spend $100 to improve honey bee health?

Let's imagine that the government, in an unprecedented move, has decided to run a 'Help the bees' campaign as a kind of social experiment. They give each of us $100 to spend on improving honey bee health. How would you spend that money? After nearly two years of writing this book and being immersed in the world of honey bee health reports and research, here is how I'd spend mine.

The first $40 would go towards funding research into organisations that are producing safe, effective *Varroa* control methods. We desperately need new ways of controlling *Varroa*. These parasitic mites and the deformed wing virus that they spread represent perhaps the biggest threat to honey bee health. If you don't treat your colonies for *Varroa*, they will eventually die by this parasite or the pathogens it spreads. Control methods might involve the rearing of resistant queens or even biological control agents. New approaches that involve genetically modifying the bacteria that live in the gut of the honey bee deserve consideration too.[23] A commercial beekeeper I work with spends NZ$50,000 each year on *Varroa* control. That money could be saved by the use of bees with bacteria modified to control these parasites, and/or with queens that display resistance.

My next $30 would go towards disease management and control. Specifically, if we really want to eliminate American foulbrood from managed beehives, let's appropriately fund the work needed to achieve it. We need epidemiological models, efficient detection tools, and boots on the ground. Elimination is a great aim – though, after looking at American foulbrood eradication programmes in places like Jersey Island, I'm not entirely convinced (yet) that it is possible. Nevertheless, an intense programme for American foulbrood control could have broader benefits. Perhaps this work would enable us to better understand hive distributions and densities, helping us to manage them better and reduce the risk of starvation.

As of June 2019 there are an estimated 924,973 registered beehives in New Zealand. We have never had so many registered hives in the history of beekeeping. There is tremendous interest from hobby beekeepers. *Photo: Phil Lester*

My last $30 would to go towards surveillance. To manage bee health well in any country, I think a multi-pronged surveillance programme is essential. This would include installing sentinel hives at ports of entry, enabling the early detection of arriving pests such as the small hive beetle or the Asian hornet. Research in the United Kingdom has demonstrated that sentinel hives can have a 'profound' impact on the early detection of honey bee pathogens and parasites, with even a small number of sentinel apiaries helping to prevent large-scale undetected outbreaks.[24] Sentinel hives placed throughout a country could also be used to monitor pesticides in pollen, and pathogen abundance and effects, and a standardised disease management system would allow for high-quality comparative data on hive survival. We could think outside the box for some cost-effective options for this surveillance – perhaps beekeeping clubs could operate these sentinel hives. The data would enable governments to make data-driven decisions about pesticide regulation and use. More broadly, sentinel hives could give us insight into insect biodiversity loss. Finally, given that starvation is such an issue in hive mortality, we need to better manage hive densities and the broader environments that provide resources for bees.

Sadly, the chances of the government giving each of us some money to spend on bee health is extremely low. The honey bee industry in New Zealand hasn't come to the party either. In 2019, New Zealand commercial beekeepers defeated (resoundingly, with three-quarters of votes against) an attempt by Apiculture New Zealand to introduce a commodity levy. Perhaps the timing of the vote was bad, after a poor season that caused financial stress on commercial beekeepers.[25] Others have viewed the situation less favourably, describing the fractured industry as 'a basketcase'.[26] Financial stress and fractured industries aside, the health of honey bees in countries such as New Zealand will only be improved by an injection of resources and a clear set of priorities or goals. In order to achieve substantive progress, it will be vital for the industry to coalesce around goals, objectives and a methodology to get there.

Everyone would have different ideas and priorities for honey bee health. We need to have an informed discussion, and develop an unbiased, evidence-based and well-consulted ranking of risk factors – an idea which is not new. As Steinhauer et al. conclude in their 2018 paper 'Drivers of colony losses', continued surveillance will help us to monitor any changes in exposure risk. Combined with studies that reflect real-life conditions of honey bees, surveillance will enable researchers and policy makers to decide which risk factors to prioritise in their mediation efforts.[27]

The future: Humans with pots of pollen, or robot pollinators?

Every year in Hanyuan County, Sichuan, a mass pollination event occurs. This county describes itself as the 'world's pear capital'. But this pollination event doesn't involve bees. Instead, farmers hand-pollinate the flowers on their pear trees. People carry pots of pollen and paintbrushes on long sticks to reach

A farmer pollinates a pear tree by hand in the province of Sichuan, 2016. Heavy pesticide use on fruit trees in the area caused a severe decline in wild bee populations, and trees are now pollinated by hand in order to produce better fruit. *Photo: Kevin Frayer / Getty*

flowers. Children are encouraged to climb to the tops of the pear trees so they can reach the highest flowers.

Why have the natural pollinators in Hanyuan County been decimated? A lack of natural habitat and, probably, excessive pesticide use. Four decades of 8–10 pesticide sprays per season is thought to be a major factor in the decline of pollinators. Many areas are now sprawling monocultural landscapes with few opportunities for bees to collect pollen or nectar. In areas where flowering and natural vegetation have been well preserved, pollinators are abundant and the pollination of crops like apples occurs naturally, which eliminates the need for hand pollination.[28]

A dearth of pollinators could be addressed by other means. Right now, scientists around the globe are developing little robots or drones for artificial pollination. One Japanese researcher, Eijiro Miyako, has designed an insect-sized drone coated with a patch of horse hair bristles and a liquid ionic gel. The drone is able to zoom around between flowers, collecting and transferring pollen as it goes. Miyako and his team believe this technology 'has the potential in the long term to lead to a breakthrough for a sustainable society'.[29] Others view such technology with scepticism. 'There are 1 million acres of almond trees in California,' says Marla Spivak, a professor in entomology at the University of Minnesota. 'Every flower needs to be pollinated to set the nut. Two million colonies of bees are trucked in to pollinate the almonds, and each colony has between ten and twenty thousand foragers. How many robots would be needed?'[30]

Other commentators echo these views – and my own. 'On top of more practical arguments, such as costs to smaller farms, I would not like to live in a world where bees are replaced by plastic machines,' says entomologist Quinn McFrederick, of the University of California, Riverside. 'Let's focus on protecting the biodiversity we still have left.'[31]

Hand pollination in the Sichuan province, 2016. *Photo: Kevin Frayer / Getty*

Acknowledgements

I've really enjoyed writing this book and talking with many beekeepers and scientists around the world. Many people have contributed their time, expertise or photographs that have added a lot to this work. Thanks to you all!

Many thanks to the large group of people who have read or commented on chapters or sections of this book, including James Baty, Madeleine Beekman, Oksana Borowik, Mariana Bulgarella, Jana Dobelmann, Julia Eloff, Antoine Felden, John Haywood, Sarah Lester, Frank Lindsay, Benjamin Oldroyd, John MacKay and Emily Remnant. I am extremely fortunate to have had Ashleigh Young from Victoria University of Wellington Press edit this book. Thanks, Ashleigh.

Notes

Introduction

1 N. Gallai et al., 'Economic valuation of the vulnerability of world agriculture confronted with pollinator decline', *Ecol. Econ.* 68, no.3 (2009): 810–21, doi: 10.1016/j.ecolecon.2008.06.014

2 A.M. Klein et al., 'Importance of pollinators in changing landscapes for world crops', *Proc. Royal Soc. B* 274, no.1608 (2007): 303–13, doi: 10.1098/rspb.2006.3721

3 R. Abcarian, 'How the honey bee crisis is affecting California's almond growers', *Los Angeles Times*, 26 Feb 2016, latimes.com/local/abcarian/la-me-abcarian-bees-almonds-20160226-column.html

4 J. Eilperin, 'How the White House plans to help the humble bee maintain its buzz', *Washington Post*, 19 May 2015, washingtonpost.com/politics/whats-all-the-obama-buzz-about-bees/2015/05/18/5ebd1580-fd6a-11e4-805c-c3f407e5a9e9_story.html

5 K. Kulhanek et al., 'A national survey of managed honey bee 2015–2016 annual colony losses in the USA', *J. Apic. Res.* 56, no.4 (2017): 328–40, doi: 10.1080/00218839.2017.1344496

6 A. Skerrett, '"Love our bees": Annual Bee Awareness Month kicks off', *Newshub*, 29 Aug 2019, newshub.co.nz/home/rural/2019/08/love-our-bees-annual-bee-awareness-month-kicks-off.html

7 C. Burgess and C. Dubbs, *Animals in Space: From Research Rockets to the Space Shuttle* (Basel, Switzerland: Springer Nature, 2007).

8 T.E. Nelson and J.R. Peterson, 'Experiment results: Insect flight observation at zero gravity', NASA / Houston University (1982): 68, ntrs.nasa.gov/archive/nasa/casi.ntrs.nasa.gov/19830025642.pdf

9 D. Poskevich, 'A comparison of honeycomb structures built by *Apis mellifera*', *Shuttle Student Involvement Program (SSIP) Final Reports of Experiments flown* (SE82-17) (1989).

10 G. Fleming, *Animal Plagues: Their History, Nature, and Prevention* (Piccadilly, UK: Chapman and Hall, 1871).

11 T. Short, *A General Chronological History of the Air, Weather, Seasons, Meteors, &c. in Sundry Places and Different Times, volume 1* (T. Longman and A. Miller: London, UK, 1749), 144.

12 G. Waldbauer, *Millions of Monarchs, Bunches of Beetles: How Insects Find Strength in Numbers* (Cambridge, MA: Harvard University Press, 2000).

13 L. Bailey, 'The "Isle of Wight disease": The origin and significance of the myth', *Bee World* 45 (1963): 1–9, doi: 10.1080/0005772X.1964.11097032

14 D. vanEngelsdorp et al., 'Colony Collapse Disorder: A descriptive study', *PLoS ONE* 4, no.8 (2009): e6481, doi: 10.1371/journal.pone.0006481

15 D. Cox-Foster and D. vanEngelsdorp, 'Solving the mystery of the vanishing bees', *Sci. Am.* 300, no.4 (2009): 40–47, doi: 10.1038/scientificamerican0409-40, scientificamerican.com/article/saving-the-honeybee/

16 Ibid.

17 Kulhanek et al., 'A national survey'.

18 Fleming, *Animal Plagues*, xxix.

19 J.R. de Miranda, G. Cordoni and G. Budge, 'The acute bee paralysis virus–Kashmir bee virus–Israeli acute paralysis virus complex', *J. Invertebr. Pathol.* 103 (2010): S33, doi: 10.1016/j.jip.2009.06.014

20 Manaaki Whenua – Landcare Research, '2018 Bee colony loss survey', landcareresearch.co.nz/science/portfolios/enhancing-policy-effectiveness/bee-health

21 P. Brown and T. Robertson, *Report on the 2018 New Zealand colony loss survey* (Wellington: Landcare Research, 2019), mpi.govt.nz/dmsdocument/33663/direct

22 S. Hogan, 'Beekeepers say there's not enough food for all hives as more and more are created', 1 News, 3 Feb 2019, tvnz.co.nz/one-news/new-zealand/beekeepers-say-theres-not-enough-food-all-hives-more-and-created

23 F. Sánchez-Bayo and K.A.G. Wyckhuys, 'Worldwide decline of the entomofauna: A review of its drivers', *Biol. Conserv.* 232 (2019): 10, doi: 10.1016/j.biocon.2019.01.020

24 D. Goulson et al., 'Bee declines driven by combined stress from parasites, pesticides, and lack of flowers', *Science* 347, no.6229 (2015): 1255957-2, doi: 10.1126/science.1255957

25 D. Carrington, 'Plummeting insect numbers "threaten collapse of nature"', *Guardian*, 10 Feb 2019, theguardian.com/environment/2019/feb/10/plummeting-insect-numbers-threaten-collapse-of-nature

26 S. Palmer, 'Insectagedden: New Zealanders have "two weeks of life" after insect apocalype – expert', Newshub, 12 Feb 2019, newshub.co.nz/home/new-zealand/2019/02/insectageddon-new-zealanders-have-two-weeks-of-life-after-insect-apocalypse-expert.html

27 P.J. Lester et al., 'The long-term population dynamics of common wasps in their native and invaded range', *J. Anim. Ecol.* 86, no.2 (2017): 337–47, doi: 10.1111/1365-2656.12622

28 M. Brown, 'Disappearing insects cause for concern', *Marlborough Express*, 17 Jan 2019, stuff.co.nz/environment/109865556/disappearing-insects-cause-for-concern

29 Cox-Foster and vanEngelsdorp, 'Solving the mystery of the vanishing bees', 46.

30 E.A.D. Mitchell et al., 'A worldwide survey of neonicotinoids in honey', *Science* 358, no.6359 (2017): 109–11, doi: 10.1126/science.aan3684

31 P. Barkham, 'The evidence is clear: insecticides kill bees. The industry denials look

absurd', *Guardian*, 3 July 2017, theguardian.com/commentisfree/2017/jul/03/insecticides-kill-bees-denials-absurd

32 B.A. Woodcock et al., 'Country-specific effects of neonicotinoid pesticides on honey bees and wild bees', *Science* 356 (2017): 1393–95, doi: 10.1126/science.aaa1190

1. The life history of bees

1 Jürgen Tautz, *The Buzz About Bees: Biology of a Superorganism* (Berlin, Germany: Springer-Verlag Berlin Heidelberg, 2008), 3.

2 M. Nagari et al., 'Signals from the brood modulate the sleep of brood tending bumblebee workers', *bioRxiv* (2018), doi: 10.1101/500744

3 H. Zwaka et al., 'Context odor presentation during sleep enhances memory in honeybees', *Curr. Biol.* 25, no.1 (2015): 2869–874, doi: 10.1016/j.cub.2015.09.069

4 B.A. Klein et al., 'Sleep deprivation impairs precision of waggle dance signaling in honey bees', *PNAS* 107, no.52 (2010): 22705–709, doi: 10.1073/pnas.1009439108

5 B.B. Land and T.D. Seeley, 'The grooming invitation dance of the honey bee', *Ethology* 110, no.1 (2004): 1–10, doi: 10.1046/j.1439-0310.2003.00947.x

6 M. Veits et al., 'Flowers respond to pollinator sound within minutes by increasing nectar sugar concentration', *Ecol. Lett.* 22, no.9 (2019): 1483–89, doi: 10.1111/ele.13331

7 E. Yong, 'Plants can hear animals using their flowers', *Atlantic*, 10 Jan 2019, theatlantic.com/science/archive/2019/01/plants-use-flowers-hear-buzz-animals/579964/

8 K.E. Anderson et al., 'Hive-stored pollen of honey bees: Many lines of evidence are consistent with pollen preservation, not nutrient conversion', *Mol. Ecol.* 23, no.23 (2014): 5904–17, doi: 10.1111/mec.12966

9 H.N. Scofield and H.R. Mattila, 'Honey bee workers that are pollen stressed as larvae become poor foragers and waggle dancers as adults', *PLoS ONE* 10, no.4 (2015): e0121731, doi: 10.1371/journal.pone.0121731

10 M. Simone-Finstrom and M. Spivak, 'Propolis and bee health: The natural history and significance of resin use by honey bees', *Apidologie* 41 (2010): 295–311, doi: 10.1051/apido/2010016

11 Ibid.

12 M. Kieliszek et al., 'Pollen and bee bread as new health-oriented products: A review', *Trends Food Sci. Tech.* 71 (2018): 170–80, doi: 10.1016/j.tifs.2017.10.021

13 V.R. Pasupuleti et al., 'Honey, propolis, and royal jelly: A comprehensive review of their biological actions and health benefits', *Oxid. Med. Cell. Longev.* (2017): 1259510, doi: 10.1155/2017/1259510

14 M. Nouvian, J. Reinhard and M. Giurfa, 'The defensive response of the honeybee *Apis mellifera*', *J. Exp. Biol.* 219 (2016): 3505–17, doi: 10.1242/jeb.143016

15 T. Fujiyuki et al., 'Novel insect picorna-like virus identified in the brains of aggressive worker honeybees', *J. Virol.* 78, no.3 (2004): 1093–100, doi: 10.1128/jvi.78.3.1093-1100.2004

16 M.D. Breed, E. Guzman-Novoa and G.J. Hunt, 'Defensive behavior of honey bees: organization, genetics, and comparisons with other bees', *Ann. Rev. Entomol.* 49 (2004): 271–98, doi: 10.1146/annurev.ento.49.061802.123155

17 J.O. Schmidt, *The Sting of the Wild* (John Hopkins University Press, 2016), 226.

18 E. Yong, 'The worst places to get stung by a bee: Nostril, lip, penis', *National Geographic*, 3 April 2014, nationalgeographic.com/science/phenomena/2014/04/03/the-worst-places-to-get-stung-by-a-bee-nostril-lip-penis/

19 R.E. Welton, D.J. Williams and D. Liew, 'Injury trends from envenoming in Australia, 2000–2013', *Intern. Med. J.* 47, no.2 (2017): 170–76, doi: 10.1111/imj.13297

20 R. Burlew, 'What are winter bees and what do they do?', *Honey Bee Suite*, 2017, honeybeesuite.com/what-are-winter-bees-and-what-do-they-do/

21 M. Rich, 'Japan's working mothers: record responsibilities, little help from dads', *New York Times*, 2 Feb 2019, nytimes.com/2019/02/02/world/asia/japan-working-mothers.html

22 T.D. Seeley, 'The effect of drone comb on a honey bee colony's production of honey', *Apidologie* 33, no.1 (2002): 75–86, doi: 10.1051/apido:2001008

23 T.S.K. Johansson and M.P. Johansson, 'Effects of drone comb on brood and honey production in honey bee colonies', *Ann. Entomol. Soc. Am.* 64, no.4 (1971): 954–56, doi: 10.1093/aesa/64.4.954

24 E. Baudry et al., 'Relatedness among honeybees (*Apis mellifera*) of a drone congregation', *Proc. Royal Soc. B* 265, no.1409 (1998): 2009–14, doi: 10.1098/rspb.1998.0533

25 J.M. Withrow and D.R. Tarpy, 'Cryptic "royal" subfamilies in honey bee (*Apis mellifera*) colonies', *PLoS ONE* 13, no.7 (2018): e0199124, doi: 10.1371/journal.pone.0199124

26 S.S. Schneider, 'The honey bee colony: Life history', in *The hive and the honey bee*, ed. J.M. Graham (Hamilton, IL: Dadant and Sons, 2015), 73–109.

27 K. Weintraub, 'A rare bird indeed: A cardinal that's half male, half female', *New York Times*, 9 Feb 2019, nytimes.com/2019/02/09/science/cardinal-sex-gender.html

28 S.E. Aamidor et al., 'Sex mosaics in the honeybee: How haplodiploidy makes possible the evolution of novel forms of reproduction in social Hymenoptera', *Biol. Lett.* 14, no.11 (2018): 20180670-4, doi: 10.1098/rsbl.2018.0670

29 Ibid.

30 D.R. Tarpy and D.J.C. Fletcher, '"Spraying" behavior during queen competition in honey bees', *J. Insect Behav.* 16 (2003): 425–37, doi: 10.1023/A:1024884211098

31 T.D. Seeley, *Honeybee Democracy* (Princeton, NJ: Princeton University Press, 2011), 273.

32 Ibid, 236.

2. *Varroa destructor*: The vampire mite

1 'Case Study 4: Response to the incursion of the *Varroa* bee mite', in *Report of the Controller and Auditor-General – Management of Biosecurity Risks: Case Studies* (Wellington, 2002), 81, oag. parliament.nz/2002/biosecurity-case-studies/docs/part4.pdf

2 H. Benard et al., 'The outbreak of *Varroa destructor* in New Zealand bees: Delimiting survey results and management options', *Surveillance Magazine* 28, no.3 (2001): 3–5, sciquest. org.nz/node/47198

3 'Case Study 4', 81.

4 A. Fricker, 'Plight of the humble bee', *New Zealand Geographic* 66 (2003): 102–11, nzgeo. com/stories/plight-of-the-humble-bee/

5 'Case Study 4', 85.

6 Ibid, 88.

7 F. Mondet et al., 'On the front line: Quantitative virus dynamics in honeybee (*Apis mellifera* L.) colonies along a new expansion front of the parasite *Varroa destructor*', *PLoS Pathog.* 10, no.8 (2014): e1004323, doi: 10.1371/journal.ppat.1004323

8 P. Rosenkranz, P. Aumeier and B. Ziegelmann, 'Biology and control of *Varroa destructor*', *J. Invertebr. Pathol.* 103 (2010): S96–S119, doi: 10.1016/j.jip.2009.07.016

9 A.C. Oudemans, *The Great Sea Serpent* (Leiden: E.J. Brill, London: Luzac & co, 1892).

10 G.M. Eberhart, *Mysterious Creatures: A Guide to Cryptozoology* (ABC-CLIO, 2002), 477.

11 A.C. Oudemans, 'On a new genus and species of parasitic Acari', *Notes from the Leyden Museum* 24 (1904): 216–22, biodiversitylibrary.org/page/9644749#page/238/mode/1up

12 Ibid, 220.

13 D. De Jong, R.A. Morse and G.C. Eickwort, 'Mite pests of honey bees', *Annu. Rev. Entomol.* 27 (1982): 230, doi: 10.1146/annurev.en.27.010182.001305

14 Ibid, 231.

15 Ibid, 231.

16 N. Bakalar, 'Earth may be home to a trillion species of microbes', *New York Times*, 24 May 2016, nytimes.com/2016/05/24/science/one-trillion-microbes-on-earth.html

17 D.L. Anderson and J.W.H. Trueman, '*Varroa jacobsoni* (Acari: Varroidae) is more than one species', *Exp. Appl. Acarol.* 24 (2000): 165, doi: 10.1023/A:1006456720416

18 D. Pattemore, 'What's up with bee numbers in New Zealand?', *New Zealand Herald*, 28

Sept 2016, nzherald.co.nz/nz/news/article.cfm?c_id=1&objectid=11715582

19 Rosenkranz, Aumeier and Ziegelmann, S96.

20 S.J. Martin, 'Ontogenesis of the mite *Varroa jacobsoni* Oud. in worker brood of the honeybee *Apis mellifera* L. under natural conditions', *Exp. Appl. Acarol.* 18 (1994): 87–100, doi: 10.1007/BF00055033

21 S.J. Martin, 'Ontogenesis of the mite *Varroa jacobsoni* Oud. in drone brood of the honeybee *Apis mellifera* L. under natural conditions', *Exp. Appl. Acarol.* 19 (1995): 199–210, doi: 10.1007/BF00130823

22 A. Easton, 'Intestinal worms impair child health in the Philippines', *Br. Med. J.* 318 (1999): 214, doi: 10.1136/bmj.318.7178.214

23 P. Deavoll, 'Fighting the good fight for varroa resistance in bees', Stuff, 11 March 2016, stuff.co.nz/business/farming/77730773/fighting-the-good-fight-for-varroa-resistance-in-bees

24 S.D. Ramsey et al., '*Varroa destructor* feeds primarily on honey bee fat body tissue and not hemolymph', *Proc. Natl. Acad. Sci. USA* 116, no.5 (2019): 1792, doi: 10.1073/pnas.1818371116

25 Ibid, 1795.

26 G. Di Prisco et al., 'A mutualistic symbiosis between a parasitic mite and a pathogenic virus undermines honey bee immunity and health', *Proc. Natl. Acad. Sci. USA* 113, no.2 (2016): 3203–208, doi: 10.1073/pnas.1523515113

27 P.G. Kevan et al., 'A summary of the Varroa-virus disease complex in honey bees', *Am. Bee J.* 146 (2006): 694–97, atrium.lib.uoguelph.ca/xmlui/handle/10214/2413

28 'Varroa mites and associated honey bee diseases more severe than previously thought', *Entomology Today*, 27 April 2016, entomologytoday.org/2016/04/27/varroa-mites-and-associated-honey-bee-diseases-more-severe-than-previously-thought/

29 A. Matheson and M. Reid, *Practical beekeeping in New Zealand*, 4th ed. (Auckland: Exisle Publishing, 2011), 285.

30 J. Wegener et al., 'Pathogenesis of varroosis at the level of the honey bee (*Apis mellifera*) colony', *J. Insect Physiol.* 91–92 (2016): 1–9, doi: 10.1016/j.jinsphys.2016.06.004

31 M. Goodwin and M. Taylor, *Control of Varroa: A guide for New Zealand beekeepers* (New Zealand Ministry of Agriculture and Forestry, 2007), 182.

32 Deavoll, 'Fighting the good fight'.

33 Y. Garbian et al., 'Bidirectional transfer of RNAi between honey bee and *Varroa destructor*: *Varroa* gene silencing reduces *Varroa* population', *PLoS Pathog.* 8 (2012): e1003035, doi: 10.1371/journal.ppat.1003035

34 S.P. Leonard et al., 'Engineered symbionts activate honey bee immunity and limit pathogens', *Science* 367, no.6477 (2020): 573–76, doi: 10.1126/science.aax9039

35 Ibid.

36 R.J. Paxton, 'A microbiome silver bullet for honey bees', *Science* 367, no.6477 (2020): 504–06, doi: 10.1126/science.aba6135

37 B.A. Harpur et al., 'Integrative genomics reveals the genetics and evolution of the honey bee's social immune system', *Genome Biol. Evol.* 11, no.3 (2019): 937–48, doi: 10.1093/gbe/evz018

38 F. Mondet et al., 'Antennae hold a key to *Varroa*-sensitive hygiene behaviour in honey bees', *Sci. Rep.* 5, no.10454 (2015), doi: 10.1038/srep10454

39 J. Carroll, 'No regional development cash to breed the "perfect" varroa-resistant honey bee', Stuff, 27 July 2018, stuff.co.nz/business/farming/105582644/no-regional-development-cash-to-breed-the-perfect-varroaresistant-honey-bee

40 Rosenkranz, Aumeier and Ziegelmann, 'Biology and control of *Varroa destructor*', S107.

41 M. Oddie et al. 'Rapid parallel evolution overcomes global honey bee parasite', *Sci. Rep.* 8, no.7704 (2018), doi: 10.1038/s41598-018-26001-7

42 Ibid, 6.

43 G.J. Mordecai et al., 'Superinfection exclusion and the long-term survival of honey bees in Varroa-infested colonies', *ISME J.* 10 (2016): 1182–191.

44 'Mystery of the "superbees" solved, Vita Bee Health, 2 Nov 2015, vita-europe.com/beehealth/blog/mystery-of-the-superbees-solved/

45 Rosenkranz, Aumeier and Ziegelmann, 'Biology and control of *Varroa destructor*', S112.

3. Viruses

1 C. Arnold, *Pandemic 1918: Eyewitness accounts from the greatest medical holocaust in modern history* (St. Martin's Press, 2018).

2 'Influenza hits Samoa', NZHistory, Ministry for Culture and Heritage, 2020, nzhistory.govt.nz/media/photo/influenza-pandemic-hits-samoa

3 L. Wilfert et al., 'Deformed wing virus is a recent global epidemic in honeybees driven by *Varroa* mites', *Science* 351, no.6273 (2016): 594–97, doi: 10.1126/science.aac9976

4 F.L. Horsfall, *Thomas Milton Rivers: 1888–1962* (Washington, DC: National Academy of Sciences, 1965), 270, nasonline.org/publications/biographical-memoirs/memoir-pdfs/rivers-thomas.pdf

5 L.P. Villarreal, 'Are viruses alive?', *Sci. Am.* 291, no.6 (2004), scientificamerican.com/article/are-viruses-alive-2004/

6 Ibid.

7 M.A.M. Gruber et al., 'Single-stranded RNA viruses infecting the invasive Argentine

ant, *Linepithema humile*', *Sci. Rep.* 7, no.3304 (2017): 1–10, doi: 10.1038/s41598-017-03508-z

8 R. Andino and E. Domingo, 'Viral quasispecies', *Virology* 479–80 (2015): 46–51, doi: 10.1016/j.virol.2015.03.022

9 A.J. McMenamin and M.L. Flenniken, 'Recently identified bee viruses and their impact on bee pollinators', *Curr. Opin. Insect. Sci.* 26 (2018): 120, doi: 10.1016/j.cois.2018.02.009

10 E.J. Remnant et al., 'A diverse range of novel RNA viruses in geographically distinct honey bee populations', *J. Virol.* 91, no.16 (2017): 1, doi: 10.1128/JVI.00158-17

11 C.M. Grozinger and M.L. Flenniken, 'Bee viruses: Ecology, pathogenicity, and impacts', *Annu. Rev. Entomol.* 64 (2019): 205–26, doi: 10.1146/annurev-ento-011118-111942

12 M.J. Roossinck, 'The good viruses: Viral mutualistic symbioses', *Nat. Rev. Microbiol* 9, no.2 (2011): 99–108, doi: 10.1038/nrmicro2491

13 Wilfert et al., 'Deformed wing virus is a recent global epidemic', 594.

14 M.A. Furst et al., 'Disease associations between honeybees and bumblebees as a threat to wild pollinators', *Nature* 506, no.7488 (2014): 364–66, doi: 10.1038/nature12977

15 D.T. Peck, M.L. Smith and T.D. Seeley, '*Varroa destructor* mites can nimbly climb from flowers onto foraging honey bees', *PLoS ONE* 1, no.12 (2016): e0167798, doi: 10.1371/journal.pone.0167798

16 D.C. Schroeder and S.J. Martin, 'Deformed wing virus: The main suspect in unexplained honeybee deaths worldwide', *Virulence* 3, no.7 (2012): 589–91, doi: 10.4161/viru.22219

17 J.R. de Miranda and E. Genersch, 'Deformed wing virus', *J. Invertebr. Pathol.* 103 (2010): S48–S61, doi: 10.1016/j.jip.2009.06.012

18 J. Iqbal and U. Mueller, 'Virus infection causes specific learning deficits in honeybee foragers', *Proc. Royal Soc. B* 274 (2007): 1517–521, doi: 10.1098/rspb.2007.0022

19 J. Hubert et al., 'Changes in the bacteriome of honey bees associated with the parasite *Varroa destructor*, and pathogens *Nosema* and *Lotmaria passim*', *Microb. Ecol.* 73, no.3 (2017): 685–98, doi: 10.1007/s00248-016-0869-7

20 de Miranda and Genersch, 'Deformed wing virus', S52.

21 K. Benaets et al., 'Covert deformed wing virus infections have long-term deleterious effects on honeybee foraging and survival', *Proc. Royal Soc. B* 284 (2017): 20162149, doi: 10.1098/rspb.2016.2149

22 S.J. Martin et al., 'Global honey bee viral landscape altered by a parasitic mite', *Science* 336, no.6086 (2012): 1304–306, doi: 10.1126/science.1220941

23 M.E. Natsopoulou et al., 'The virulent, emerging genotype B of deformed wing virus is closely linked to overwinter honeybee worker loss', *Sci. Rep.* 7, no.5242 (2017), doi: 10.1038/s41598-017-05596-3

24 Schroeder and Martin, 'Deformed wing virus: The main suspect', 589.

25 D. Fears, 'Scientists say spread of virus that's killing honeybees "a man-made thing"', *Washington Post*, 7 Feb 2016, chicagotribune.com/nation-world/ct-honeybees-virus-humans-20160207-story.html

26 Wilfert et al., 'Deformed wing virus is a recent global epidemic', 595.

27 E.J. Remnant et al., 'Direct transmission by injection affects competition among RNA viruses in honeybees', *Proc. Royal Soc. B* 286, no.1895 (2019): 20182452, doi: 10.1098/rspb.2018.2452

28 'Study shows dangerous bee virus might be innocent bystander', Phys.org, Jan 2019, phys.org/news/2019-01-dangerous-bee-virus-innocent-bystander.html

29 S. Gisder and E. Genersch, 'Viruses of commercialized insect pollinators', *J. Invertebr. Pathol.* 147 (2017): 51–59, doi: 10.1016/j.jip.2016.07.010

30 Gruber et al., 'Single-stranded RNA viruses', 1.

31 Gisder and Genersch, 'Viruses of commercialized insect pollinators', 52.

32 L. Bailey and R.D. Woods, 'Two more small RNA viruses from honey bees and further observations on sacbrood and acute bee-paralysis viruses', *J. Gen. Virol.* 37 (1977): 179, doi: org/10.1099/0022-1317-37-1-175

33 J.H. Todd, J.R. de Miranda and B.V. Ball, 'Incidence and molecular characterization of viruses found in dying New Zealand honey bee (*Apis mellifera*) colonies infested with *Varroa destructor*', *Apidologie* 38, no.4 (2007): 354–55, doi: 10.1051/apido:2007021

34 D. Cox-Foster and D. vanEngelsdorp, 'Solving the mystery of the vanishing bees', *Sci. Am.* 300, no.4 (2009): 40–47, scientificamerican.com/article/saving-the-honeybee/

35 Y.P. Chen et al., 'Israeli acute paralysis virus: epidemiology, pathogenesis and implications for honey bee health', *PLoS Pathog.* 10, no.7 (2014): 2–3, doi: 10.1371/journal.ppat.1004261

36 A.M.J. McFadden et al., 'Israeli acute paralysis virus not detected in *Apis mellifera* in New Zealand in a national survey', *J. Apic. Res.* 53, no.5 (2014): 520–27, doi: 10.3896/ibra.1.53.5.03

37 G. Foulsham, '10 million viruses in one drop of seawater', Futurity, 15 Aug 2011, futurity.org/millions-of-marine-viruses-ebb-and-flow/

38 Remnant et al., 'A diverse range of novel RNA viruses', 1.

39 R.S. Cornman et al., 'Pathogen webs in collapsing honey bee colonies', *PLoS ONE* 7, no.7 (2012): e43562, doi: 10.1371/journal.pone.0043562

40 G.E. Budge et al., 'Chronic bee paralysis as a serious emerging threat to honey bees', *Nat. Commun.* 11 (2020): 2164, doi: 10.1038/s41467-020-15919-0

41 M. Ribière, Violaine Olivier and Philippe Blanchard, 'Chronic bee paralysis: A disease and a virus like no other?', *J. Inverter. Pathol.* 103 (2010): S121, doi: 10.1016/j.jip.2009.06.013

42 L. Bailey, 'The "Isle of Wight disease": The origin and significance of the myth', *Bee World* 45 (1963).

43 Ibid.

44 G. Di Prisco et al., 'Neonicotinoid clothianidin adversely affects insect immunity and promotes replication of a viral pathogen in honey bees', *Proc. Natl. Acad. Sci. USA* 110, no.46 (2013): 18466–471, doi: 10.1073/pnas.1314923110

45 Grozinger and Flenniken, 'Bee viruses: Ecology, pathogenicity, and impacts', 207–8.

46 G. DeGrandi-Hoffman and Y. Chen, 'Nutrition, immunity and viral infections in honey bees', *Curr. Opin. Insect. Sci.* 10 (2015): 170, doi: 10.1016/j.cois.2015.05.007

47 G. DeGrandi-Hoffman et al., 'The effect of diet on protein concentration, hypopharyngeal gland development and virus load in worker honey bees (*Apis mellifera* L.)', *J. Insect Physiol.* 56, no.9 (2010): 1184–191, doi: 10.1016/j.jinsphys.2010.03.017

48 A.G. Dolezal et al., 'Interacting stressors matter: diet quality and virus infection in honeybee health', *R. Soc. Open Sci.* 6, no.2 (2019): 181803, doi: 10.1098/rsos.181803

49 treesforbeesnz.org (2019)

50 A.J. McMenamin and E. Genersch, 'Honey bee colony losses and associated viruses', *Curr. Opin. Insect. Sci.* 8 (2015): 121–29, doi: 10.1016/j.cois.2015.01.015

51 P.E. Stamets et al., 'Extracts of polypore mushroom mycelia reduce viruses in honey bees', *Sci. Rep.* 8 (2018): 13936, doi: 10.1038/s41598-018-32194-8

52 J. Flynn Mogensen, 'A new study shows how mushrooms could save bees. Yes, mushrooms', *Mother Jones*, 4 Oct 2018, motherjones.com/environment/2018/10/a-new-study-shows-how-mushrooms-could-save-bees-yes-mushrooms/

4. American foulbroud

1 Aristotle, *The complete works of Aristotle, volume 1: The revised Oxford translation*, ed. Jonathan Barnes (Princeton, NJ: Princeton University Press, 2014), 974.

2 D. Coleman, 'Entertaining entomology: Insects and insect performers in the eighteenth century', *Eighteenth-Century Life* 30 (2006): 107–34, doi: 10.1215/00982601-2006-004

3 F. Maderspacher, 'All the queen's men', *Curr. Biol.* 17 (2007): R191–95, doi: 10.1016/j.cub.2007.02.017

4 A.G. Schirach, *Sächsischer bienenvater* (Adam Jacob Spiekermann: Leipzig and Zittau, 1766), 637–41.

5 C. Ash, F.G. Priest, and M.D. Collins, 'Molecular identification of rRNA group 3 bacilli (Ash, Farrow, Wallbanks and Collins) using a PCR probe test. Proposal for the creation of a new genus *Paenibacillus*', *Antonie Van Leeuwenhoek* 64 (1993): 253–60, doi: 10.1007/BF00873085

6 E. Genersch, 'American Foulbrood in honey bees and its causative agent, *Paenibacillus*

larvae', *J. Invertebr. Pathol.* 103 Suppl 1 (2010): S10–19, doi: 10.1016/j.jip.2009.06.015

7 E. Genersch, 'Foulbrood diseases of honey bees – from science to practice', in *Beekeeping – From Science to Practice*, ed. R.H. Vreeland and D. Sammataro (Springer International Publishing, 2017), 159.

8 R.J. Cano and M.K. Borucki, 'Revival and identification of bacterial spores in 25- to 40-million-year-old Dominican amber', *Science* 268, no.5213 (1995): 1060–064, doi: 10.1126/science.7538699

9 H. Shimanuki and D.A. Knox, 'Susceptibility of *Bacillus larvae* to terramycin', *Am. Bee J.* 134, no.2 (1994): 125–26.

10 Genersch, 'American Foulbrood in honey bees', S10–19.

11 C.J. Brødsgaard, W. Ritter and H. Hansen, 'Response of in vitro reared honey bee larvae to various doses of *Paenibacillus larvae* larvae spores', *Apidologie* 29 (1998): 569–78.

12 T.R. Hoage and W.C. Rothenbuhler, 'Larval honey bee response to various doses of *Bacillus larvae* spores', *J. Econ. Entomol.* 59 (1966): 42–45, doi: 10.1093/jee/59.1.42

13 Genersch, 'Foulbreed disease of honey bees', 170.

14 Ibid, 159.

15 Genersch, 'American Foulbrood in honey bees', S10–19.

16 H. Beimsr et al., 'Discovery of *Paenibacillus larvae* ERIC V: Phenotypic and genomic comparison to genotypes ERIC I-IV reveal different inventories of virulence factors which correlate with epidemiological prevalences of American Foulbrood', *Int. J. Med. Microbiol.* 310, no.2 (2020): 151394, doi: 10.1016/j.ijmm.2020.151394

17 Genersch, 'Foulbrood diseases of honey bees', 163.

18 Ibid. 165–66.

19 Genersch, 'American Foulbrood in honeybees', S10–19.

20 M.O. Schafer et al., 'Rapid identification of differentially virulent genotypes of *Paenibacillus larvae*, the causative organism of American foulbrood of honey bees, by whole cell MALDI-TOF mass spectrometry', *Vet. Microbiol.* 170 (2014): 291–97, doi: 10.1016/j.vetmic.2014.02.006

21 Y. Ueno et al., 'Population structure and antimicrobial susceptibility of *Paenibacillus larvae* isolates from American foulbrood cases in *Apis mellifera* in Japan', *Environ. Microbiol. Rep.* 10 (2018): 210–16, doi: 10.1111/1758-2229.12623

22 Schafer et al., 'Rapid identification of differentially virulent genotypes', 291–97.

23 S.A.M. Graham, 'American foulbrood and its causative agent, *Paenibacillus larvae*, in New Zealand's registered hives and apiaries' (Master of Science thesis, Victoria University of Wellington, 2014), researcharchive.vuw.ac.nz/handle/10063/4801

24 E. Genersch, A. Ashiralieva and I. Fries, 'Strain- and genotype-specific differences

in virulence of *Paenibacillus larvae* subsp. *larvae*, a bacterial pathogen causing American foulbrood disease in honeybees', *Appl. Environ. Microbiol.* 71 (2005): 7551–55, doi: 10.1128/AEM.71.11.7551-7555.2005

25 A. Lindström, S. Korpela and I. Fries, 'Horizontal transmission of *Paenibacillus larvae* spores between honey bee (*Apis mellifera*) colonies through robbing', *Apidologie* 39 (2008): 515–22, doi: 10.1051/apido:2008032 (2008).

26 R.M. Goodwin, J.H. Perry and A.T. Houten, 'The effect of drifting honey bees on the spread of American foul brood infections', *J. Apic. Res.* 33 (1994): 209–12.

27 Ibid.

28 M.A. Hornitzky, 'The spread of *Paenibacillus larvae* subsp *larvae* infections in an apiary', *J. Apic. Res.* 37 (1998): 261–65.

29 Lindström, Korpela and Fries, 'Horizontal transmission', 515–22.

30 Genersch, 'Foulbrood diseases of honey bees', 157–74.

31 M. Goodwin, *Elimination of American foulbrood disease without the use of drugs: A practical manual for beekeepers* (National Beekeepers' Association of New Zealand, 2006), 27.

32 I. Fries, A. Lindström and S. Korpela, 'Vertical transmission of American foulbrood (*Paenibacillus larvae*) in honey bees (*Apis mellifera*)', *Vet. Microbiol.* 114 (2006): 269–74, doi: 10.1016/j.vetmic.2005.11.068

33 L. Stowell, 'Foulbrood: Screaming bees distress keeper', *NZ Herald*, 8 June 2017, nzherald.co.nz/nz/news/article.cfm?c_id=1&objectid=11870524

34 Management Agency, National American Foulbrood Pest Management Plan New Zealand, 'Burning AFB colonies', n.d., afb.org.nz/burning-afb-colonies/

35 University of Warwick, 'Model of dangerous bee disease in Jersey provides tool in fight against honeybee infections', Phys.org, 16 Sept 2013, phys.org/news/2013-09-dangerous-bee-disease-jersey-tool.html

36 S. Datta et al., 'Modelling the spread of American foulbrood in honeybees', *J. R. Soc. Interface* 10 (2013): 20130650, doi: 10.1098/rsif.2013.0650

37 'Bee keepers warned about American Foulbrood', Gov.je, 21 May 2019, gov.je/News/2019/Pages/AmericanFoulbrood2019.aspx

38 Genersch, 'Foulbrood diseases of honey bees', 166.

39 Hobbler, post to 'General Beekeeping', NZ Beekeepers, 7 Feb 2012, nzbees.net/forums/topic/584-shook-swarming-in-new-zealand/

40 W. Reybroeck et al., 'Antimicrobials in beekeeping', *Vet. Microbiol.* 158 (2012): 1–11, doi: 10.1016/j.vetmic.2012.01.012

41 H.W. Smith, M.B. Huggins, and K.M. Shaw, 'The control of experimental *Escherichia coli* diarrhoea in calves by means of bacteriophages', *J. Gen. Microbiol.* 133, no.5 (1987): 1111–126.

42 S. Ghorbani-Nezami et al., 'Phage therapy is effective in protecting honeybee larvae from American foulbrood disease', *J. Insect Sci.* 15 (2015): 1–5, doi: 10.1093/jisesa/iev051

43 J. Morton, 'Virus vs disease: New bid to help our honeybees', *NZ Herald*, 15 Jan 2018, nzherald.co.nz/nz/news/article.cfm?c_id=1&objectid=11975301

44 H. Salmela, G.V. Amdam and D. Freitak, 'Transfer of immunity from mother to offspring is mediated via egg-yolk protein vitellogenin', *PLoS Pathog.* 11 (2015): e1005015, doi: 10.1371/journal.ppat.1005015

45 B. Locke, M. Low and E. Forsgren, 'An integrated management strategy to prevent outbreaks and eliminate infection pressure of American foulbrood disease in a commercial beekeeping operation', *Prev. Vet. Med.* 167 (2019): 48–52, doi: 10.1016/j.prevetmed.2019.03.023

46 Goodwin, *Elimination of American foulbrood disease*, 9.

47 M. Goodwin, 'American foulbrood control: The New Zealand approach', *Bee World* 86 (2005): 44–45.

48 C. King, 'AFB Pest Management Plan: Challenges and constraints' (paper presented at the Apiculture Conference, Rotorua Energy Events Centre, Rotorua, June 2019), apicultureconference2019.co.nz/wp-content/uploads/2019/08/1100-AFB-PMP-Challenges-and-Constraints_Clifton-King.pdf

49 Lindström, Korpela and Fries, 'Horizontal transmission', 515–22.

50 R.M. Goodwin, A.T. Houten and J.H. Perry, 'Incidence of American foulbrood infections in feral honey bee colonies in New Zealand', *New Zeal. J. Zool.* 21 (1994): 285–87.

51 A-L. Woolf, 'Predator Free 2050 "dream" called too costly, too unlikely at Biological Heritage conference', *Dominion Post*, 21 May 2019, stuff.co.nz/environment/112879315/predator-free-2050-dream-called-too-costly-too-unlikely-at-biological-heritage-conference

5. Pathogens

1 R. Tipa, 'Bee colonies wiped out as new parasite spreads through New Zealand', Stuff, 22 June 2015, stuff.co.nz/business/farming/agribusiness/69531572/

2 W.J. Sutherland, P. Barnard, S, Broad et al., 'A 2017 horizon scan of emerging issues for global conservation and biological diversity', *Trends Ecol. Evol.* 32, no.1 (2017): 31–40, doi: 10.1016/j.tree.2016.11.005

3 Wikipedia, s.v. 'Trypanosomatida', last modified 22 May 2020, en.wikipedia.org/wiki/Trypanosomatida

4 G. Poinar, Jr., '*Lutzomyia adiketis* sp. n. (Diptera: Phlebotomidae), a vector of *Paleoleishmania neotropicum* sp. n. (Kinetoplastida: Trypanosomatidae) in Dominican amber', *Parasit. Vectors*

1, no.22 (2008), doi: 10.1186/1756-3305-1-22

5 M.J.F. Brown, R. Schmid-Hempel and P. Schmid-Hempel, 'Strong context-dependent virulence in a host-parasite system: reconciling genetic evidence with theory', *J. Anim. Ecol.* 72 (2003): 999, doi: 10.1046/j.1365-2656.2003.00770.x

6 R.J. Gegear, M.C. Otterstatter, and J.D. Thomson, 'Does parasitic infection impair the ability of bumblebees to learn flower-handling techniques?', *Anim. Behav.* 70, no.1 (2005): 209–15, doi: 10.1016/j.anbehav.2004.09.025

7 M.C. Otterstatter et al., 'Effects of parasitic mites and protozoa on the flower constancy and foraging rate of bumble bees', *Behav. Ecol. Sociobiol.* 58 (2005): 383–89, doi: 10.1007/s00265-005-0945-3

8 D. Goulson, *Bumblebees: Behaviour, ecology, and conservation*, 2nd ed (Oxford University Press, 2012), 317.

9 D.F. Langridge and R.B. McGhee, '*Crithidia mellificae* n. sp. an acidophilic Trypanosomatid of the honey bee *Apis mellifera*', *J. Eukaryot. Microbiol.* 14, no.3 (1967): 485–87, doi: 10.1111/j.1550-7408.1967.tb02033.x

10 M.J.F. Brown, R. Loosli and P. Schmid-Hempel, 'Condition-dependent expression of virulence in a trypanosome infecting bumblebees', *Oikos* 91, no.3 (2000): 421, doi: 10.1034/j.1600-0706.2000.910302.x

11 G.M. LoCascio et al., 'Pollen from multiple sunflower cultivars and species reduces a common bumblebee gut pathogen', *R. Soc. Open Sci.* 6 (2019): 190279, doi: 10.1098/rsos.190279

12 E.C. Palmer-Young et al., 'Bumble bee parasite strains vary in resistance to phytochemicals', *Sci. Rep.* 6, no.37087 (2016), doi: 10.1038/srep37087

13 LoCascio et al., 'Pollen from multiple sunflower cultivars and species', 7.

14 H. Koch and P. Schmid-Hempel, 'Socially transmitted gut microbiota protect bumble bees against an intestinal parasite', *Proc. Natl. Acad. Sci. USA* 108, no.48 (2011), 19288–292, doi: 10.1073/pnas.1110474108

15 B.K. Mockler et al., 'Microbiome structure influences infection by the parasite Crithidia bombi in bumble bees', *Appl. Environ. Microbiol.* 84 (2018): e02335-02317, doi: 10.1128/AEM.02335-17

16 B. Imhoof and P. Schmid-Hempel, 'Patterns of local adaptation of a protozoan parasite to its bumblebee host', *Oikos* 82, no.1 (1998): 63, doi: 10.2307/3546917

17 D. Goulson et al., 'Bee declines driven by combined stress from parasites, pesticides, and lack of flowers', *Science* 347, no.6229 (2015): 1255957, doi: 10.1126/science.1255957

18 H.B. Fantham and A. Porter, 'The morphology, biology and economic importance of *Nosema bombi*, N. sp., parasitic in various humble bees (*Bombus* spp.)', *Ann. Trop. Med. Parasitol.* 8, no.3 (1914): 623–38, doi: 10.1080/00034983.1914.11687667

19 O. Otti and P. Schmid-Hempel, '*Nosema bombi*: A pollinator parasite with detrimental fitness effects', *J. Invertebr. Pathol.* 96 (2007): 120, doi: 10.1016/j.jip.2007.03.016

20 B. Imhoof and P. Schmid-Hempel, 'Colony success of the bumble bee, *Bombus terrestris*, in relation to infections by two protozoan parasites, *Crithidia bombi* and *Nosema bombi*', *Insectes Soc.* 46 (1999): 233–38, doi: 10.1007/s000400050139

21 Goulson, *Bumblebees: Behaviour, ecology, and conservation*, 317.

22 D. Goulson, G.C. Lye and B. Darvill, 'Decline and conservation of bumble bees', *Annu. Rev. Entomol.* 53 (2008): 191–208, doi: 10.1146/annurev.ento.53.103106.093454

23 S.A. Cameron et al., 'Patterns of widespread decline in North American bumble bees', *Proc. Natl. Acad. Sci. U.S.A.* 108, no.2 (2011), 662–67, doi: 10.1073/pnas.1014743108

24 R. Schmid-Hempel et al., 'The invasion of southern South America by imported bumblebees and associated parasites', *J. Anim. Ecol.* 83 (2014): 823–37, doi: 10.1111/1365-2656.12185

25 M.A. Aizen et al., 'Coordinated species importation policies are needed to reduce serious invasions globally: The case of alien bumblebees in South America', *J. Appl. Ecol.* 56 (2018): 100–06, doi: 10.1111/1365-2664.13121

26 Sutherland et al., 'A 2017 horizon scan', 31–40.

27 Aizen, Smith-Ramírez and Morales, 'Coordinated species importation policies', 100–06.

28 L. Pasteur, 'Études sur la maladie des vers à soie: moyen pratique assuré de la combattre et d'en prévenir le retour' (Gauthier-Villars, successeur de Mallet-Bachelier, 1870).

29 I. Fries, '*Nosema apis* – A parasite in the honey bee colony', *Bee World* 74, no.1 (1993): 5–19, doi: 10.1080/0005772X.1993.11099149

30 J.D. Ellis and P.A. Munn, 'The worldwide health status of honey bees', *Bee World* 86, no.4 (2005): 88–101, doi: 10.1080/0005772x.2005.11417323

31 'The "Isle of Wight" bee disease', *Nature* 99 (1917): 508, doi: 10.1038/099507a0

32 R. Martín-Hernández, C. Bartolomé, N. Chejanovsky et al., '*Nosema ceranae* in *Apis mellifera*: A 12 years postdetection perspective', *Environ. Microbiol.* 20 (2018): 1313, doi: 10.1111/1462-2920.14103

33 W. Li, Y. Chen and S.C. Cook, 'Chronic *Nosema ceranae* infection inflicts comprehensive and persistent immunosuppression and accelerated lipid loss in host *Apis mellifera* honey bees', *Int. J. Parasitol.* 48, no.6 (2018): 433–44, doi: 10.1016/j.ijpara.2017.11.004

34 I. Fries et al., '*Nosema ceranae* n. sp. (Microspora, Nosematidae), morphological and molecular characterization of a microsporidian parasite of the Asian honey bee *Apis cerana* (Hymenoptera, Apidae)', *Eur. J. Protistol.* 32 (1996): 356–365, doi: 10.1016/s0932-4739(96)80059-9

35 S. Plischuk et al., 'South American native bumblebees (Hymenoptera: Apidae) infected

by *Nosema ceranae* (*Microsporidia*), an emerging pathogen of honeybees (*Apis mellifera*)', *Environ. Microbiol. Rep.* 1, no.2 (2009): 131–35, doi: 10.1111/j.1758-2229.2009.00018.x

36 E.W. Teixeira et al., '*Nosema ceranae* has been present in Brazil for more than three decades infecting Africanized honey bees', *J. Invertebr. Pathol.* 114, no.3 (2013): 250–54, doi: 10.1016/j.jip.2013.09.002

37 I. Fries, '*Nosema ceranae* in European honey bees (*Apis mellifera*)', *J. Invertebr. Pathol.* 103 Suppl 1 (2010): S74, doi: 10.1016/j.jip.2009.06.017

38 Z.L. Murray and P.J. Lester, 'Confirmation of *Nosema ceranae* in New Zealand and a phylogenetic comparison of *Nosema* spp. Strains', *J. Apic. Res.* 54, no.2 (2016): 101–04, doi: 10.1080/00218839.2015.1101240

39 A. Vilcinskas et al., 'Invasive harlequin ladybird carries biological weapons against native competitors', *Science* 340 (2013): 862–63, doi: 10.1126/science.1234032

40 Fries, '*Nosema ceranae* in European honey bees', S74.

41 Martín-Hernández et al., '*Nosema ceranae* in *Apis mellifera*', 1315.

42 Ibid, 1314–315.

43 R.J. Paxton, 'Does infection by *Nosema ceranae* cause "Colony Collapse Disorder" in honey bees (*Apis mellifera*)?', *J. Apic. Res.* 49 (2015): 80–84, doi: 10.3896/ibra.1.49.1.11

44 L. Pasteur, 'Études sur la maladie des vers à soie', 179–206.

45 Q. Huang et al., 'Survival and immune response of drones of a Nosemosis tolerant honey bee strain towards *N. ceranae* infections', *J. Invertebr. Pathol.* 109, no.3 (2012): 297–302, doi: 10.1016/j.jip.2012.01.004

46 V. Rada et al., 'Microflora in the honeybee digestive tract: counts, characteristics and sensitivity to veterinary drugs', *Apidologie* 28 (1997): 357–65, doi: 10.1051/apido:19970603

47 J.P. van den Heever et al., 'Evaluation of Fumagilin-B® and other potential alternative chemotherapies against *Nosema ceranae*-infected honeybees (*Apis mellifera*) in cage trial assays', *Apidologie* 47 (2015): 617–30, doi: 10.1007/s13592-015-0409-3

48 W.F. Huang et al., '*Nosema ceranae* escapes fumagillin control in honey bees', *PLoS Pathog.* 9, no.3 (2013): e1003185, doi: 10.1371/journal.ppat.1003185

49 A.J. Burnham, 'Scientific advances in controlling *Nosema ceranae* (*Microsporidia*) infections in honey bees (*Apis mellifera*)', *Front. Vet. Sci.* 6 (2019): 79, doi: 10.3389/fvets.2019.00079

50 C. Rodríguez-García et al., 'Nosemosis control in European honey bees, *Apis mellifera*, by silencing the gene encoding *Nosema ceranae* polar tube protein 3', *J. Exp. Biol.* 221 (2018), doi: 10.1242/jeb.184606

51 W.J. Fuchs, *Lotmaria passim: Ruth Lotmar, bee researcher and zoologist* (Digiboo, Küsnacht, Switzerland, 2019), 54.

52 R.S. Schwarz et al., 'Characterization of two species of Trypanosomatidae from the

honey bee *Apis mellifera: Crithidia mellificae* Langridge and McGhee, and *Lotmaria passim* n. gen., n. sp.', *J. Eukaryot. Microbiol.* 62, no.5 (2015): 580, doi: 10.1111/jeu.12209

53 Langridge and McGhee, '*Crithidia mellificae* n. sp. an acidophilic Trypanosomatid of the honey bee *Apis mellifera*'.

54 V. Strobl et al., 'Trypanosomatid parasites infecting managed honeybees and wild solitary bees', *Int. J. Parasitol.* 49 (2019): 605–13, doi: 10.1016/j.ijpara.2019.03.006

55 J. Ravoet et al., 'Comprehensive bee pathogen screening in Belgium reveals *Crithidia mellificae* as a new contributory factor to winter mortality', *PLoS ONE* 8, no.8 (2013): e72443, doi: 10.1371/journal.pone.0072443

56 C. Runckel et al., 'Temporal analysis of the honey bee microbiome reveals four novel viruses and seasonal prevalence of known viruses, *Nosema*, and *Crithidia*', *PLoS ONE* 6, no.6 (2011): e20656, doi: 10.1371/journal.pone.0020656

57 R.S. Cornman et al., 'Pathogen webs in collapsing honey bee colonies', *PLoS ONE* 7, no.8 (2012): e43562-5, doi: 10.1371/journal.pone.0043562

58 Schwarz et al., 'Characterization of two species of Trypanosomatidae', 580.

59 L.L. Richardson et al., 'Secondary metabolites in floral nectar reduce parasite infections in bumblebees', *Proc. R. Soc. B* 282, no.1803 (2015): 20142471, doi: 10.1098/rspb.2014.2471

60 K. de Sousa Pereira et al., 'Double-stranded RNA reduces growth rates of the gut parasite *Crithidia mellificae*', *Parasitol. Res.* 118 (2019): 715–21, doi: 10.1007/s00436-018-6176-0

61 Cornman et al., 'Pathogen webs in collapsing honey bee colonies', e43562-1.

62 R.S. Schwarz and J.D. Evans, 'Single and mixed-species trypanosome and microsporidia infections elicit distinct, ephemeral cellular and humoral immune responses in honey bees', *Dev. Comp. Immunol.* 40 (2013): 300–10, doi: 10.1016/j.dci.2013.03.010

6. Pesticides

1 M. Luxmoore, 'Rural Russia abuzz as beekeepers rally to thwart pesticide use', RFE/RL, 26 July 2019, rferl.org/a/rural-russia-abuzz-as-beekeepers-rally-to-thwart-pesticide-use/30077516.html

2 E. Gershkovich, 'What's killing Russia's honey bees?', *Moscow Times*, 26 July 2019, themoscowtimes.com/2019/07/26/what-killing-russia-honey-bees-a66563

3 Ibid.

4 S. Savary et al., 'The global burden of pathogens and pests on major food crops', *Nat. Ecol. Evol.* 3 (2019): 430–39, doi: 10.1038/s41559-018-0793-y

5 C.A. Mullin et al., 'High levels of miticides and agrochemicals in North American

apiaries: implications for honey bee health', *PLoS ONE* 5, no.3 (2010): e9754, doi: 10.1371/journal.pone.0009754

6 D. vanEngelsdorp et al., 'Colony Collapse Disorder: A descriptive study', *PLoS ONE* 4, no.8 (2009): e6481, doi: 10.1371/journal.pone.0006481

7 Mullin et al., 'High levels of miticides and agrochemicals', 1.

8 D. Cox-Foster and D. vanEngelsdorp, 'Solving the mystery of the vanishing bees', *Sci. Am.* 300, no.4 (2009): 40–47, scientificamerican.com/article/saving-the-honeybee/

9 Mullin et al., 'High levels of miticides and agrochemicals', 17.

10 J. Bernal et al., 'Overview of pesticide residues in stored pollen and their potential effect on bee colony (*Apis mellifera*) losses in Spain', *J. Econ. Entomol.* 103, no.6 (2010): 1964–71, doi: 10.1603/ec10235

11 vanEngelsdorp et al., 'Colony Collapse Disorder', 12.

12 D. van Engelsdorp et al., 'A survey of honey bee colony losses in the U.S., fall 2007 to spring 2008', *PLoS ONE* 3, no.12 (2008): 4, doi: 10.1371/journal.pone.0004071

13 T. Reilly et al., 'Occurrence of boscalid and other selected fungicides in surface water and groundwater in three targeted use areas in the United States', *Chemosphere* 89, no.3 (2012): 228–34, doi: 10.1016/j.chemosphere.2012.04.023

14 N. Tsvetkov et al., 'Chronic exposure to neonicotinoids reduces honey bee health near corn crops', *Science* 356, no.6345 (2017): 1395–397, doi: 10.1126/science.aam7470

15 E.D. Pilling and P.C. Jepson, 'Synergism between EBI fungicides and a pyrethroid insecticide in the honeybee (*Apis mellifera*)', *Pest Manag. Sci.* 39 (1993): 293–97, doi: 10.1002/ps.2780390407

16 N. Cedergreen, 'Quantifying synergy: A systematic review of mixture toxicity studies within environmental toxicology', *PLoS ONE* 9 (2014): e96580, doi: 10.1371/journal.pone.0096580

17 A. Iverson et al., 'Synergistic effects of three sterol biosynthesis inhibiting fungicides on the toxicity of a pyrethroid and neonicotinoid insecticide to bumble bees', *Apidologie* 50 (2019): 733–44, doi: 10.1007/s13592-019-00681-0

18 M. Henry et al., 'A common pesticide decreases foraging success and survival in honey bees', *Science* 336, no.6079 (2012): 348–50, doi: 10.1126/science.1215039

19 E.A.D. Mitchell et al., 'A worldwide survey of neonicotinoids in honey', *Science* 358, no.6359 (2017): 109–11, doi: 10.1126/science.aan3684

20 L.A. Morandin et al., 'Lethal and sub-lethal effects of spinosad on bumble bees (*Bombus impatiens* Cresson)', *Pest Manag. Sci.* 61, no.7 (2005): 619–26, doi: 10.1002/ps.1058

21 W.F. Barbosa et al., 'Biopesticide-induced behavioral and morphological alterations in the stingless bee *Melipona quadrifasciata*', *Environ. Toxicol. Chem.* 34, no.9 (2015): 2149–58, doi: 10.1002/etc.3053

22 E.V.S. Motta, K. Raymann and N.A. Moran, 'Glyphosate perturbs the gut microbiota of honey bees', *Proc. Natl. Acad. Sci. U.S.A.* 115 (2018): 10305–310, doi: 10.1073/pnas.1803880115

23 M.R. Faita et al., 'Changes in hypopharyngeal glands of nurse bees (*Apis mellifera*) induced by pollen-containing sublethal doses of the herbicide Roundup (R)', *Chemosphere* 211 (2018): 566–72, doi: 10.1016/j.chemosphere.2018.07.189

24 C. Hawes et al., 'Responses of plants and invertebrate trophic groups to contrasting herbicide regimes in the Farm Scale Evaluations of genetically modified herbicide-tolerant crops', *Philos. Trans. R. Soc. Lond., B, Biol. Sci.* 358, no.1439 (2003): 1899–913, doi: 10.1098/rstb.2003.1406

25 C. Vidau et al., 'Exposure to sublethal doses of fipronil and thiacloprid highly increases mortality of honeybees previously infected by *Nosema ceranae*', *PLoS ONE* 6 (2011): e21550, doi: 10.1371/journal.pone.0021550

26 J.S. Pettis et al., 'Crop pollination exposes honey bees to pesticides which alters their susceptibility to the gut pathogen *Nosema ceranae*', *PLoS ONE* 8 (2013): e70182, doi: 10.1371/journal.pone.0070182

27 B. Locke et al., 'Acaricide treatment affects viral dynamics in *Varroa destructor*-infested honey bee colonies via both host physiology and mite control', *Appl. Environ. Microbiol.* 78, no.1 (2012): 227–35, doi: 10.1128/AEM.06094-11

28 J. López et al., 'Sublethal pesticide doses negatively affect survival and the cellular responses in American foulbrood-infected honeybee larvae', *Sci. Rep.* 7 (2017): 40853, doi: 10.1038/srep40853

29 S.T. O'Neal, T.D. Anderson and J.Y. Wu-Smart, 'Interactions between pesticides and pathogen susceptibility in honey bees', *Curr. Opin. Insect Sci.* 26 (2018): 57–62, doi: 10.1016/j.cois.2018.01.006

30 E. Collison et al., 'Interactive effects of pesticide exposure and pathogen infection on bee health – A critical analysis', *Biol. Rev. Camb. Philos. Soc.* 91 (2016): 1006–19, doi: 10.1111/brv.12206

31 T.J Wood and D. Goulson, 'The environmental risks of neonicotinoid pesticides: A review of the evidence post 2013', *Environ. Sci. Pollut. Res.* 24 (2017): 17305, doi: 10.1007/s11356-017-9240-x

32 M. Henry et al., 'Reconciling laboratory and field assessments of neonicotinoid toxicity to honeybees', *Proc. R. Soc. B* 282, no.1819 (2015): 20152110, doi: 10.1098/rspb.2015.2110

33 B.A. Woodcock et al., 'Country-specific effects of neonicotinoid pesticides on honey bees and wild bees', *Science* 356 (2017): 1393–95.

34 Ibid.

35 M. Rundlöf et al., 'Seed coating with a neonicotinoid insecticide negatively affects wild bees', *Nature* 521 (2015): 9, doi: 10.1038/nature14420

36 'Expert reaction to two new papers on bees and neonicotinoids', Science Media Centre, 22 April 2015, sciencemediacentre.org/expert-reaction-to-two-new-papers-on-bees-and-neonicotinoids/

37 J. Osterman et al., 'Clothianidin seed-treatment has no detectable negative impact on honeybee colonies and their pathogens', *Nat. Commun.* 10, no.692 (2019): 1, doi: 10.1038/s41467-019-08523-4

38 Henry et al., 'Reconciling laboratory and field assessments', 20152110.

39 Osterman et al., 'Clothianidin seed-treatment', 9.

40 N. Steinhauer et al., 'Drivers of colony losses', *Curr. Opin. Insect Sci.* 26 (2018): 144, doi: 10.1016/j.cois.2018.02.004

41 M.L. Eng, B.J.M. Stutchbury and C.A. Morrissey, 'A neonicotinoid insecticide reduces fueling and delays migration in songbirds', *Science* 365, no.6458 (2019): 1177–80, doi: 10.1126/science.aaw9419

42 M. Yamamuro et al., 'Neonicotinoids disrupt aquatic food webs and decrease fishery yields', *Science* 366, no.6465 (2019): 620–23, doi: 10.1126/science.aax3442

43 N.L. Carreck, 'A beekeeper's perspective on the neonicotinoid ban', *Pest Manag. Sci.* 73 (2017): 1295–98, doi: 10.1002/ps.4489

44 S. Tosi and J.C. Nieh, 'Lethal and sublethal synergistic effects of a new systemic pesticide, flupyradifurone (Sivanto), on honeybees', *Proc. R. Soc. B* 286 (2019): 20190433, doi: 10.1098/rspb.2019.0433

45 'Government rejects vote to ban synthetic pesticides in Switzerland', *Le News*, 26 June 2019, lenews.ch/2019/06/26/government-rejects-vote-to-ban-synthetic-pesticides-in-switzerland/

46 P. Brown et al., 'Winter 2016 honey bee colony losses in New Zealand', *J. Apic. Res.* 57, no.2 (2018): 278–91, doi: 10.1080/00218839.2018.1430980

47 D. MacLeod, 'Science and Research Focus Group report', *New Zealand Beekeeper* (Dec 2018): 9–11.

48 E. Urlacher et al., 'Measurements of chlorpyrifos levels in forager bees and comparison with levels that disrupt honey bee odor-mediated learning under laboratory conditions', *J. Chem. Ecol.* 42 (2016): 127–38, doi: 10.1007/s10886-016-0672-4

49 J.C. Hartle et al., 'Chemical contaminants in raw and pasteurized human milk', *J. Hum. Lact.* 34 (2018): 340–49, doi: 10.1177/0890334418759308

7. Predators

1 S. McKenzie, 'Winnie-the-Pooh's skull goes on display, shows tooth decay from honey',

CNN, 20 Nov 2015, edition.cnn.com/2015/11/20/europe/winnie-the-pooh-skull-honey-teeth-london/index.html

2. P. Brown and T. Robertson, *Report on the 2018 New Zealand colony loss survey* (Wellington: Landcare Research, 2019), mpi.govt.nz/dmsdocument/33663/direct

3 P.J. Lester and J.R. Beggs, 'Invasion success and management strategies for social *Vespula* wasps', *Annu. Rev. Entomol.* 64 (2019): 51–71, doi: 10.1146/annurev-ento-011118-111812

4 P.J. Lester, *The vulgar wasp: The story of a ruthless invader and ingenious predator* (Wellington: Victoria University Press, 2018), 74.

5 Lester and Beggs, 'Invasion success and management strategies', 62–63.

6 D. Laurino et al., '*Vespa velutina*: An alien driver of honey bee colony losses', *Diversity* 12, no.1 (2019): 1, doi: 10.3390/d12010005

7 Ibid, 4.

8 N. Vigdor, 'Asian giant hornet invasion threatens honey bees in Pacific Northwest', *New York Times*, 24 Dec 2019, nytimes.com/2019/12/24/us/asian-giant-hornet.html

9 S.D. McClain, 'Asian giant hornet, a nemesis of honeybees, appears in Washington', Capital Press, 2020, capitalpress.com/ag_sectors/orchards_nuts_vines/asian-giant-hornet-a-nemesis-of-honeybees-appears-in-washington/article_773e5a90-2dc3-11ea-b587-9b98cdd79a23.html

10 T. Giraud, J.S. Pedersen and L. Keller, 'Evolution of supercolonies: The Argentine ants of southern Europe', *Proc. Natl. Acad. Sci. U.S.A.* 99 (2002): 6075–79, doi: 10.1073/pnas.092694199

11 E. Sunamura et al., 'Intercontinental union of Argentine ants: behavioral relationships among introduced populations in Europe, North America, and Asia', *Insectes Soc.* 56 (2009): 143–47, doi: 10.1007/s00040-009-0001-9

12 Brown and Robertson, *Report on the 2018 New Zealand colony loss survey.*

13 M.A.M. Gruber et al., 'Single-stranded RNA viruses infecting the invasive Argentine ant, *Linepithema humile*, *Sci. Rep.* 7 (2017): 3304, doi: 10.1038/s41598-017-03508-z

14 R.J. Deslippe and W.D. Melvin, 'Assessment of ant foraging on beehives in an apiary infested with *Solenopsis invicta* (Hymenoptera : Formicidae)', *Southwest. Entomol.* 26 (2001): 215–19.

15 Ibid.

16 A.N. Payne, T.F. Shepherd and J. Rangel, 'The detection of honey bee (*Apis mellifera*)-associated viruses in ants', *Sci. Rep.* 10 (2020): 2923, doi: 10.1038/s41598-020-59712-x

17 W.M. Hood, P.M. Horton and J.W. McCreadie, 'Field evaluation of the red imported fire ant (Hymenoptera: Formicidae) for the control of wax moths (Lepidoptera: Pyralidae) in stored honey bee comb', *J. Agric. Urban Entomol.* 20, no.2 (2003): 93–103.

18 'Yucatan beekeepers in crisis', *Yucatan Times*, 19 March 2019, theyucatantimes.com/2019/03/yucatan-beekeepers-in-crisis/

19 S. Mohamadzade Namin et al., 'Invasion pathway of the honeybee pest, small hive beetle, *Aethina tumida* (Coleoptera: Nitidulidae) in the Republic of Korea inferred by mitochondrial DNA sequence analysis', *J. Asia Pac. Entomol.* 22, no.3 (2019): 963–68, doi: 10.1016/j.aspen.2019.07.008

20 P. Neumann, J.S. Pettis and M.O. Schäfer, 'Quo vadis *Aethina tumida*? Biology and control of small hive beetles', *Apidologie* 47 (2016): 427–66, doi: 10.1007/s13592-016-0426-x

21 J.D. Ellis and H.R. Hepburn, 'An ecological digest of the small hive beetle (*Aethina tumida*), a symbiont in honey bee colonies (*Apis mellifera*)', *Insectes Soc.* 53 (2006): 8–19, doi: 10.1007/s00040-005-0851-8

22 Neumann, Pettis and Schäfer, 'Quo vadis *Aethina tumida*?', 442.

23 P. Neumann and P.J. Elzen, 'The biology of the small hive beetle (*Aethina tumida*, Coleoptera: Nitidulidae): Gaps in our knowledge of an invasive species', *Apidologie* 35 (2004): 229–47, doi: 10.1051/apido:2004010

24 Ellis and Hepburn, 'An ecological digest', 12.

25 F.O. Idrissou et al., 'International beeswax trade facilitates small hive beetle invasions', *Sci. Rep.* 9 (2019): 10665, doi: 10.1038/s41598-019-47107-6

26 M.O. Schäfer et al., 'How to slow the global spread of small hive beetles, *Aethina tumida*', *Biol. Invasions* 21 (2019): 1451–459, doi: 10.1007/s10530-019-01917-x

27 C.A. Kwadha et al., 'The biology and control of the greater wax moth, *Galleria mellonella*', *Insects* 8, no.2 (2017): 61, doi: 10.3390/insects8020061

28 Ibid.

29 J.D. Ellis, J.R. Graham and A. Mortensen, 'Standard methods for wax moth research', *J. Apic. Res.* 52 (2015): 1–17, doi: 10.3896/ibra.1.52.1.10

30 J.S. Pettis et al., 'A rapid survey technique for *Tropilaelaps* mite (Mesostigmata: Laelapidae) detection', *J. Econ. Entomol.* 106 (2013): 1535–44, doi: 10.1603/ec12339

31 P. Bombelli, C.J. Howe and F. Bertocchini, 'Polyethylene bio-degradation by caterpillars of the wax moth *Galleria mellonella*', *Curr. Biol.* 27 (2017): R292–R293, doi: 10.1016/j.cub.2017.02.060

32 B.J. Cassone et al., 'Role of the intestinal microbiome in low-density polyethylene degradation by caterpillar larvae of the greater wax moth, *Galleria mellonella*', *Proc. R. Soc. B* 287 (2020): 20200112, doi: 10.1098/rspb.2020.0112 (2020).

33 M. Marshall, 'The sinister moth from *Silence of the Lambs* can squeak', BBC, 5 Aug 2015, bbc.com/earth/story/20150805-terrifying-squeak-of-death-moth

34 J.S. Pettis and W.T. Wilson, 'Life history of the honey bee tracheal mite (Acari: Tarsonemidae)', *Ann. Entomol. Soc. Am.* 89 (1996): 368–74, doi: 10.1093/aesa/89.3.368

35 L. Bailey, 'The epidemiology of the infestation of the honey bee, *Apis mellifera* L., by the mite *Acarapis woodi* (Rennie) and the mortality of infested bees', *Parasitology* 48, no.3–4 (1958): 493–506, doi: 10.1017/S0031182000021430

36 L. Bailey, 'The "Isle of Wight disease": The origin and significance of the myth', *Bee World* 45, no.1 (1963): 1–9, doi: 10.1080/0005772X.1964.11097032

37 M. Frazier et al., 'The incidence and impact of honey bee tracheal mites and nosema disease on colony mortality in Pennsylvania', *Bee Sci.* 3, no.2 (1994): 94–100.

38 R. Ochoa et al., 'Observations on the honey bee tracheal mite *Acarapis woodi* (Acari: Tarsonemidae) using low-temperature scanning electron microscopy', *Exp. Appl. Acarol.* 35 (2005): 239–49, doi: 10.1007/s10493-004-5080-8

39 R.G. Danka and J.D. Villa, 'Evidence of autogrooming as a mechanism of honey bee resistance to tracheal mite infestation', *J. Apic. Res.* 37 (2015): 39–46, doi: 10.1080/00218839.1998.11100953

40 B.B. Land and T.D. Seeley, 'The grooming invitation dance of the honey bee', *Ethology* 110 (2004): 1–10, doi: 10.1046/j.1439-0310.2003.00947.x

41 J.S. Pettis and T. Pankiw, 'Grooming behavior by *Apis mellifera* L. in the presence of *Acarapis woodi* (Rennie) (Acari: Tarsonemidae)', *Apidologie* 29 (1998): 241–53, doi: 10.1051/apido:19980304

42 Y. Sakamoto et al., 'Differential susceptibility to the tracheal mite *Acarapis woodi* between *Apis cerana* and *Apis mellifera*', *Apidologie* 48 (2016): 150–58, doi: 10.1007/s13592-016-0460-8

43 A. Wroe and K. Colquhoun, *Book of Obituaries* (Profile Books: London, 2011), 185.

44 E. Forsgren, 'European foulbrood in honey bees', *J. Invertebr. Pathol.* 103 Suppl 1 (2010): S5–9, doi: 10.1016/j.jip.2009.06.016

45 O. Lewkowski and S. Erler, 'Virulence of *Melissococcus plutonius* and secondary invaders associated with European foulbrood disease of the honey bee', *MicrobiologyOpen* 8, no.3 (2019): e00649, doi: 10.1002/mbo3.649

46 S. Wilkins, M.A. Brown and A.G. Cuthbertson, 'The incidence of honey bee pests and diseases in England and Wales', *Pest Manag. Sci.* 63, no.11 (2007): 1062–68, doi: 10.1002/ps.1461

47 P. Chanpanitkitchote et al., '*Acute bee paralysis virus* occurs in the Asian honey bee *Apis cerana* and parasitic mite *Tropilaelaps mercedesae*', *J. Invertebr. Pathol.* 151 (2018): 131–36, doi: 10.1016/j.jip.2017.11.009

48 P. Chantawannakul et al., 'Tropilaelaps mite: An emerging threat to European honey bee', *Curr. Opin. Insect Sci.* 26 (2018): 69–75, doi: 10.1016/j.cois.2018.01.012

49 Ibid.

8. The future

1 S.S. Kumar, 'Colony Collapse Disorder (CCD) in honey bees caused by EMF radiation', *Bioinformation* 14 (2018): 421–24, doi: 10.6026/97320630014521

2 F. Sanchez-Bayo et al., 'Are bee diseases linked to pesticides? A brief review', *Environ. Int.* 89–90 (2016): 7–11, doi: 10.1016/j.envint.2016.01.009

3 J. Lowe, 'The super bowl of beekeeping', *New York Times*, 15 Aug 2018, nytimes.com/2018/08/15/magazine/the-super-bowl-of-beekeeping.html

4 R. Brodschneider et al., 'Multi-country loss rates of honey bee colonies during winter 2016/2017 from the COLOSS survey', *J. Apic. Res.* 57 (2018): 452–57, doi: 10.1080/00218839.2018.1460911

5 P. Brown et al., 'Winter 2016 honey bee colony losses in New Zealand', *J. Apic. Res.* 57 (2018): 278–91, doi: 10.1080/00218839.2018.1430980

6 K. Kulhanek et al., 'A national survey of managed honey bee 2015–2016 annual colony losses in the USA', *J. Apic. Res.* 56 (2017): 328–40, doi: 10.1080/00218839.2017.1344496

7 E. Amiri et al., 'Queen quality and the impact of honey bee diseases on queen health: Potential for interactions between two major threats to colony health', *Insects* 8 (2017): 2, doi: 10.3390/insects8020048

8 E. Genersch et al., 'The German bee monitoring project: a long term study to understand periodically high winter losses of honey bee colonies', *Apidologie* 41 (2010): 332–52, doi: 10.1051/apido/2010014

9 D.R. Tarpy, D. Vanengelsdorp and J.S. Pettis, 'Genetic diversity affects colony survivorship in commercial honey bee colonies', *Naturwissenschaften* 100 (2013): 723–28, doi: 10.1007/s00114-013-1065-y

10 J. Woyke, 'Natural and artificial insemination of queen honeybees', *Bee World* 43 (1962): 21–25.

11 A. Fisher and J. Rangel, 'Exposure to pesticides during development negatively affects honey bee (*Apis mellifera*) drone sperm viability', *PLoS ONE* 13 (2018): e0208630, doi: 10.1371/journal.pone.0208630

12 Amiri et al., 'Queen quality', 6.

13 D. Goulson et al., 'Bee declines driven by combined stress from parasites, pesticides, and lack of flowers', *Science* 347 (2015): 1255957, doi: 10.1126/science.1255957

14 G. May, '"Just plain stupid": Manuka honey explosion leaving Hawke's Bay bees "starving"', *Hawke's Bay Today*, 16 Jan 2019, nzherald.co.nz/business/news/article.cfm?c_id=3&objectid=12190479

15 M.D. Simone-Finstrom and M. Spivak, 'Increased resin collection after parasite challenge: A case of self-medication in honey bees?', *PLoS ONE* 7 (2012): e34601, doi:

10.1371/journal.pone.0034601

16 Goulson et al., 'Bee declines driven by combined stress'.

17 D. Goulson, 'The insect apocalypse, and why it matters', *Curr. Biol.* 29 (2019): R967–R971, doi: 10.1016/j.cub.2019.06.069

18 Manaaki Whenua – Landcare Research, '2018 Bee colony loss survey', landcareresearch.co.nz/science/portfolios/enhancing-policy-effectiveness/bee-health/2018-survey

19 G.E. Budge et al., 'Pathogens as predictors of honey bee colony strength in England and Wales', *PLoS ONE* 10 (2015): e0133228, doi: 10.1371/journal.pone.0133228

20 R.W. Currie, M. Spivak and G.S. Reuter, in *The hive and the honey bee*, ed. J.M. Graham (Hamilton, IL: Dadant & Sons, 2015), 629–70.

21 N. Steinhauer et al., 'Drivers of colony losses', *Curr. Opin. Insect Sci.* 26 (2018): 142–48, doi: 10.1016/j.cois.2018.02.004

22 D. Laurino et al., '*Vespa velutina*: An alien driver of honey bee colony losses', *Diversity* 12 (2019), doi: 10.3390/d12010005

23 S.P. Leonard et al., 'Engineered symbionts activate honey bee immunity and limit pathogens', *Science* 367 (2020): 573–76, doi: 10.1126/science.aax9039

24 M.J. Keeling et al., 'Efficient use of sentinel sites: Detection of invasive honeybee pests and diseases in the UK', *J. R. Soc. Interface* 14 (2017), doi: 10.1098/rsif.2016.0908

25 G. Hutching, 'Beekeepers vote down levy proposal by large majority', Stuff, 7 March 2019, stuff.co.nz/business/farming/111103662/beekeepers-vote-down-levy-proposal-by-large-majority

26 S. Metherell, 'NZ's 'basketcase' bee industry seeks govt help to get national body, levies', NBR, 2 April 2015, nbr.co.nz/article/nzs-basketcase-bee-industry-seeks-govt-help-get-national-body-levies-bd-170975

27 Steinhauer et al., 'Drivers of colony losses', 145.

28 U. Partap and T. Ya, 'The human pollinators of fruit crops in Maoxian County, Sichuan, China', *Mt. Res. Dev.* 32 (2012): 176–86, doi: 10.1659/mrd-journal-d-11-00108.1

29 S.A. Chechetka et al., 'Materially engineered artificial pollinators', *Chem* 2, no.2 (2017): 224–39, doi: 10.1016/j.chempr.2017.01.008

30 C. Ponti, 'Rise of the robot bees: Tiny drones turned into artificial pollinators', NPR, 3 March 2007, npr.org/sections/thesalt/2017/03/03/517785082/rise-of-the-robot-bees-tiny-drones-turned-into-artificial-pollinators

31 Ibid.

Index

Entries in **bold** denote tables; entries in *italics* denote illustrations and figures.

Australia
    freedom from *Varroa*, 62, 65
    sting mortalities in, 44
    viruses in, 99, 103
azadirachtin, *171*

bacteria
    beneficial, 97
    diversity of, 65–6
    genetically modified, 81–2
    lactic acid, 37
    see also gut bacteria
bacteriophages, 124–7, *125*
bald brood, 197
Baltimore classification system, 93
Bayer CropScience, 179, 181
Bayvarol *see* flumethrin
bee beards, 53, 55
bee bread, 37–40, 46, 51, 163
bee glue *see* propolis
beekeepers
    and colony collapse disorder, 16, 18–*21*
    competition between, 209
    compliance with American foulbrood
    control, 129–30, 133
    and Isle of Wight disease, 15, 18
    and queen problems, 19–22, 26, 206–
    208, 211
    transmitting American foulbrood, 120–1
    and *Varroa*, *20–21*, 22, 62, 64, 68, 73–74,
    80, 81, 83–86
beekeeping, future of, 30
Beekman, Madeleine, 100–1
Berenbaum, May, 189
Berry, John, 22
Big Sioux River virus, 104
biodiversity, 26, 29–30, 137, 145, 156, 183,
    216
biopesticides, 171, 174, 182–3
black queen cell virus (BQCV), 99–101,
    100, 103–4, 179

Bombus dahlbomii see Patagonian bumble bee
Bombus distinguendus see great yellow bumble
    bee
Bombus ruderatus see large garden bumble
    bee
Bombus terrestris see buff-tailed bumble bee
Borowik, Oksana, 135–8, *136*
Borucki, Monica, 111
boscalid, 167, 169–70
BQCV *see* black queen cell virus
Brandts, Charlie, *13*
brood production, 78, 114, 209
Brother Adam *see* Kehrle, Karl
Bt toxin, 17
Buckfast bees *see* Russian bees
buff-tailed bumble bee *143*, 144–6, 150
    deformed wing virus in, 98
    and pesticides, 177, 179
bumble bee (*Bombus* sp.), 24
    and buzz pollination, 37
    deformed wing virus in, 96
    exports of, 144–6, 156
    impact of pesticides on, 170, 174, 178–
    80, 184
    pathogens of, 137–44, 150
    sleep deprivation in, 34
Butov, Arnold, 158
butterflies, 24
buzz pollination, 37

Canada, 15
    American foulbrood in, 126
    bee exports to, 99, 144
    pesticides used in, 166, 169–70
Cano, Raúl, 111
captan, 167
carbaryl, 163, 167
chronic bee paralysis virus (CBPV), 104–5
colony collapse disorder (CCD), 15–19,
    135, 205
    and pathogens, 137, 149, 155–7

247

Hendrickson, Heather, 127
Henry, Mickaël, 172, 176, 179
herbicides, 28, 160
  in Canada, 170
  residues in pollen and hives, 161, 163, 167
  stress caused by, 142
  sublethal effects of, 174–5
Hinson, Eloise, *47*
hive mortality *see* colony losses
hives
    American foulbrood transfer between, 118–20
  burning, 121–4, *122*, 127
  defence of, 39–42
  impact of *Varroa* on, 77–9
    see also varoosis
  numbers of bees in, 33
  reproduction of, 56
  sitting on, 60
  as superorganisms, 31–2
  threats to, 57–8
  Varroa moving between, 74
  see also sentinel hives
homing failure, 172
honey, 14, 186
  creation of, 34–5
  neonicotinoids in, 28
honey bee (*Apis* sp.)
  aggression in, 41–2
  attacked by wasps, *187*, 188
  Crithidia and *Lotmaria* in, 153–6
  declines and collapses, 14–15, 205–6
    see also colony loss
  and global insect losses, 23–6
  hygienic behaviours of, 66–7, 83
  impact of pesticides on, 171–2, 175, 177–81
  importance of, 9–14
  improving health of, 29–30, 212–14
  life cycle of, *35*
  life forms of, 33

nosemosis in, 146–52
queens of, 51
scientific name of, 93
sex between, 48
on stamps, *25*
sting of, 43–4
and tracheal mites, 201
and *Tropilaelaps* mites, 204
and *Varroa*, 64, 66, 68, 75–7, 84
viruses in, 93–4, 104
hornet, 29, 39, 186–90, 212
Hoskins, Ron, 85–6
host odour, 70
Hungary, 177–8
hydrocarbons, 41
hygienic behaviours, 66–7, 83, 85, 114, 173
hyper-polyandry, extreme, 48
hypopharyngeal glands, 34, 149, 174

IAPV *see* Israeli acute paralysis virus
imidacloprid, 162–3, 167, 176–7
immune phenotype, extended, 141
immune system
  and antivirals, 107–8
  and bee bread, 39
  compromised or suppressed, 89, 103–4, 149, 175
  fat bodies in, 76–7
  and pesticides, 106
  stimulating, 153
  and stings, 44
Indian meal moths, *198*
influenza, 88–9
insect losses, global, 23–4
insecticides, 17, 27–8
  and bee mortality, 80, 158–9, 177, 179
  'bee safe,' 181
  in pollen and hives, 161, 167
  sublethal effects of, 174
  use of term, 160
  and wasps, 188